Published in the United States by Pen of Heart Publishing Co LLC,

pohpublishing.com

Pen of Heart and www.pohpublishing.com are registered under Pen of Heart Publishing Co LLC.

Library of Congress Cataloging-in-Publication Data has been applied for:

ISBN 978-1-961224-30-8

Ebook ISBN 978-1-961224-41-4

Book and ebook cover design by Breannia Venable

Visit us on the web,

www.pohpublishing.com

PEN OF
HEART

DEDICATION

TO THE LOVE OF MY LIFE — TORRY.
FROM THE MOMENT I MET YOU, I KNEW WE WERE MEANT TO BE. NOT JUST IN THIS LIFETIME, BUT IN EVERY ONE THAT CAME BEFORE IT,
AND EVERY ONE THAT WILL FOLLOW AFTER.
ACROSS GALAXIES, ACROSS TIME —
IT'S ALWAYS BEEN YOU.
IT ALWAYS WILL BE.

CHAPTERS

STARBORN

PROLOGUE

Dear Reader,

Do you believe the stars remember you?

Do you ever look at the night sky and feel that quiet ache – that whisper that someone, somewhere, once loved you enough to chase you across time?

They say every star that burns out is a story ending. But what they don't tell you is that endings are never final – not when love is written into the atoms of your soul.

Once, long before this world knew their names, two souls were born in the light between galaxies. They loved so fiercely the universe itself took notice – and when they fell, it wept, scattering their starlight across the void.

Each time they were reborn, the world forgot them. But they never forgot each other.

They have met as strangers, as lovers, as enemies.

They have died a thousand deaths only to find one another again.

Because some loves are not bound by time.

They are written in the stars.

And this – their story – is one of them.

With every ending, there is a beginning, my dear reader.

And every beginning starts with a single spark.

And that's where our story begins...

CHAPTER I
SIGNAL

On opposite sides of the world, two souls woke beneath the same star.

She dreamed of falling.

He dreamed of flight.

Neither knew the universe had just shifted—because of them. Somewhere between their dreams, the stars remembered their names.

Estelle Auryn

The first thing Estelle *noticed was the hum.*

Not from her phone, not from the old radiator that clanged like a ghost in winter, but from the air itself—an almost-musical vibration shivering across her ceiling, laying a lace of pale light over the chipped crown molding. She lay there and listened, half afraid to breathe, because if she exhaled too loudly the sound might vanish and take the feeling with it—the impossible, aching certainty that someone had just called her name.

Estelle.

Her own voice said it back in a whisper, and then the alarm broke the spell.

The alarm clock's buzz felt heavier that morning. She sat at the edge of her bed, heart still racing, hands trembling – the ache of a love she didn't remember. The hum was gone, but a tingle lingered in her palms as if she'd been holding a warm current. She rubbed her hands together, then reached for the journal she kept on the nightstand and drew the constellation that had followed her since childhood. There were eight points arcing like a cradle, a ninth nested at the center. As always, her lines were just slightly different from the real maps.

"Good morning, Lumen," she told the sketch without knowing why. The name felt older than language, something borrowed from another life.

She drifted to the bathroom half-awake, the remnants of

her dream still tangled in her waking thoughts. Lately, she'd been haunted by the same vision – a man made of he light. But he wasn't just light, he was warmth. Each breath of his glow seemed to thrum in her chest like a heartbeat that wasn't hers. Each morning she woke emptier, like the dream had taken something from her and left a promise in its place.

The city outside her studio was already awake, sirens far off, a bus sighing at the curb, a child laughing on the stairs. The coffee machine stuttered alive. Estelle tied her hair into a loose knot and slid into a sweater, the one with a tiny bleach stain on the cuff from a late-night experiment. She taught Intro Astrobiology *at the community college three days a week and consulted for an aerospace lab downtown the other two. On paper, her job description was modest: data wrangler, spectral line analyst, someone who could find needles in the haystack of starlight.*

Off paper, they called her in when signals misbehaved.

On the counter, next to a stack of student essays, her tablet blinked with an encrypted icon: ASTRACORPS–PRIORITY:

LUMINESCENCE FILES.

AstraCorps had fingers in every lab that studied light. Some said they funded hope. Others said they patented it. She didn't remember signing the NDA under that exact project name, but AstraCorps shuffled labels the way storms shuffled clouds. She opened it and watched a short loop: a region of the night sky flickering with low-level noise. After three seconds, a pattern emerged—two spikes, then a third, then a flare that formed a curve like a heartbeat thrown sideways. Estelle leaned closer. The flare bent into a figure-eight for a single frame, a bright knot at its center. Eight stars cradling the ninth.

Her breath fogged the tablet glass. She replayed the loop three more times. The flare came from near-Earth orbit, bounced off the ionosphere, and scattered like gold dust.

It felt...familiar.

A text from her friend Sori popped up: Coffee run? Or are you married to space again?

Estelle typed back one-handed: Space put a ring on it. Rain check. Sori was sunshine in lab form—too bright for government work, too kind for the things they studied.

She pulled on jeans, double-checked her faculty badge, and slid the journal into her bag. The hum from the ceiling was truly gone now, leaving behind only a quiet that made her ache. She touched the doorframe before leaving, the way she always did—a superstition, a promise. Today her fingers tingled again, a thin thread of warmth running into her wrist like a wire being connected. She couldn't shake the feeling that somewhere, at this exact moment, someone else had reached for the same thread.

Mykael Verran

The jet cut dawn into ribbons.

Kael rode the line between blue and black, the test aircraft's nose angling toward a sky not yet decided on day. The cockpit was more software than steel—three curved screens, a nest of instruments, a joystick that responded before he finished the thought. He didn't trust machines that knew him that well, but trust was in short supply and the pay was good, so he learned the aircraft the way he learned people: by how it behaved under stress.

“Verran, status.” The voice in his headset was clipped, the consonants shaved to metal. Director Voss did everything like it was a countdown.

“Stable,” Kael said. “Climbing past fifty thousand. Turning to course two-seven-one.”

“Begin pass. Eyes on LIDAR and flux variance.”

“Copy.”

He glanced at the talisman he'd taped to the dash—half a dog tag, its stamp sanded smooth, a thread of dark cord knotted through the hole. He didn't know why he'd kept it after the desert, didn't know why he woke up some nights with the taste of ash in his mouth and the memory of a hand in his, a hand he could never quite picture. He only knew that when he flew high enough, the noise of that memory quieted, like a storm giving up on a stubborn coast.

The instruments ticked, breathed, waited. He loved that first breath—the instant before data started talking, when the sky felt endless and clean and his body understood it *was built for this, for moving fast and trusting the line to hold. At sixty thousand feet the blue went thin. The sun tipped its bright cup over the horizon and spilled gold across the wing. Then the jet coughed—a micro-jolt more felt than heard—and all three screens rippled. A low vibration ran through his bones, the same hum that had followed him since childhood, the one he could never name.*

"Verran?" Voss snapped. "You lost vector?"

"Negative. I've got...hang on."

A bloom of static washed the readouts into white. The LIDAR painted it as vapor, then as plasma, then as nothing at all—contradictions stacked like cards. Kael swore under his breath and touched the screen. The static cleared for a heartbeat, and in that heartbeat the noise became a shape. Two spikes. A third. Then a flare, arcing into a figure-eight with a bright knot at its center.

He knew it was impossible to recognize a signal. He recognized this one anyway.

"Director, I'm seeing a transient—"

"Record and continue. Do not deviate."

The knot brightened. The instruments threw up a wall of numbers like the aircraft was trying to confess a sin in every language at once. And then, beneath the numbers, Kael heard a sound—not through the headset, not through the engine, but through the bones of his face.

A note. Low and warm. It slid into him like a memory he'd lost when he was small.

His hands steadied on the stick. He didn't realize he'd said the word out loud until the mic caught it and sent it to ground.

"Soon."

It wasn't language, not exactly. It was recognition.

"Repeat that?" Voss barked.

"Nothing," Kael said. "Transient dissipated. Resuming course."

But even after the screens returned to their obedient green, the note lingered in his teeth like a promise.

Estelle

Her first class ended with a round of shy questions and one bold request from a teenager with a galaxy pinned to her backpack: "*Do you think life out there will be like us?*"

Estelle smiled and gave the honest answer.

"*I think life out there will be like life always is—endlessly surprising.*"

As her students packed up, she caught her reflection in the whiteboard glass and almost didn't recognize herself. There was a light in her eyes that hadn't been there yesterday.

In the lab, she let herself into the side room with the card reader and dropped her bag by the console. The room smelled faintly of ozone and citrus cleanser, a combination she'd come to associate with secrets. She pulled up the LUMINESCENCE *stream and set the filters to* "*aggressive,*" *then paused. Aggressive felt wrong. She changed it to* "*gentle*" *and hit run.*

Data unspooled. Noise, noise, noise—and then the same bloom from the loop in her kitchen. The flare. The figure-eight. The center like a held breath.

She murmured, "*Where are you?*"

The console pinged. An overlay suggested a rough origin:

high atmosphere. Another overlay put a circle around a moving dot labeled TEST-47.

Her badge didn't clear the next layer of security, but Estelle understood guesses even when she couldn't prove them. Something had flown through the signal. Or the signal had reached for something that could carry it.

Her phone buzzed. Unknown number.

She let it go to voicemail and replayed the clip. The flare seemed brighter each time, like it responded to being watched.

"Don't anthropomorphize noise," her grad mentor used to chide, and Estelle would nod and then go home and sketch stars like faces. Now she dragged the clip to frame-by-frame. On the seventh frame, the center knot bulged, a petal opening. She screenshot it and traced over the shape on her tablet. Even stripped of glitter, the lines hummed. Eight points arcing like a cradle, a ninth nested at the center.

Her palm tingled again. She pressed the heel of her hand against it, like that could calm the current.

"Okay," she said to the empty room, to the signal, to the ache. "I see you."

The unknown number called back. This time she answered.

"Professor Auryn," a man said, not bothering to hide the watch in his voice. "This is Director Voss with AstraCorps. I'd like to discuss your access permissions."

Estelle leaned against the console and smiled in a way that wasn't entirely friendly. "Good afternoon, Director. Which project name are we using this week?"

A pause. "LUMINESCENCE."

"Pretty," she said. "What do you want?"

He chuckled, but it carried no warmth—just static wrapped in authority.

"We detected an anomaly at oh-six-thirty-four. Your terminal pinged the packet. We prefer our contractors observe, not...interpret."

"I prefer not to get electrocuted," she said lightly, and watched the figure-eight loop again. "I'll send my notes."

"Do that," Voss said, and cut the line.

Estelle looked at the screen a long while after the call

ended. If AstraCorps wanted to scare her off, they'd chosen the wrong week. Everything in her life felt like it had been waiting for this precise frequency, like she'd been tuning a radio in the dark and the station had finally found her.

She shut down the terminal and slid her journal from her bag. On the next page she wrote one word in the center, with a line on either side like an orbit...Soon.

Kael

The debrief room was all corners. No windows. A screen bled blue on the far wall, the kind of blue that made teeth hurt. Voss waited with a clipboard he didn't need, a habit Kael suspected he'd cultivated to remind pilots they were replaceable.

"You saw the anomaly," Voss said without preface.

"I saw a lot of things," Kael said. "Most of them were numbers."

"Humor me."

Kael considered the safe answer and chose the true one. "It felt like a signal. Structured. Not reflected trash. It had...

shape."

Voss's eyes were the color of cheap ice. "We don't assign intent to noise."

"With respect, Director, you hired me because I know the difference."

A long beat stretched between them. Kael didn't look away. He'd learned not to when men like Voss were measuring where to put the leash.

At last, Voss nodded toward the screen. The anomaly bloomed, looped, brightened. "We logged a second contact," he said. "Ground-level. Minimal. If it repeats, we'll widen the net."

Something in Kael's chest answered that, a coil tightening half a turn. "If you widen the net," he said, careful, "what do you plan to catch?"

"Patterns," Voss said. "That's all." He closed the file with a snap that said it was anything but. "Full report by end of day."

Kael left with the shape of the signal still burning behind his eyes. He didn't know that, half a world away, someone was tracing that same shape in ink.

Outside the building, the air felt warmer than it should. He looked up out of habit and saw a slice of morning sky framed by glass and steel. The same single star Estelle had watched earlier was long gone, but he had the peculiar sense of being observed anyway—not by cameras or satellites, but by something gentler, older, patient as tide. He touched the dog tag and told himself it was the afterburn of adrenaline. He told himself a lot of things that were easy to say and hard to believe.

He didn't say the word again, but it lived under his tongue like a secret: Soon.

Estelle & Kael

Etselle walked, the notebook heavy in her bag. Kael cut across town on foot from the AstraCorps campus, the day still buzzing through his nerves. One walked north, one walked south, both drawn by the same instinct neither could name: don't go inside yet. A wind moved over the water and lifted the hair at the nape of Estelle's neck. At *the same moment, an eddy of warm air slid under Kael's collar. They each stopped without meaning to, feet aligning to the same beat, faces turning toward the same patch of sky.*

The hum returned—quiet, almost tender. Streetlamps flickered. Far above, a satellite pivoted as if listening.

On the promenade, two strangers rounded the same curve from opposite ends of the path. They didn't see each other yet. Not with their eyes.

But something in the world did.

That evening, the city's lights came up like a thousand small suns. Somewhere beneath them, two signals aligned — one in the sky, one in the heart. The universe whispered the same word twice, and both of them heard it.

Soon.

CHAPTER 2
RESONANCE

Estelle

The dream began with the light again.

Not sunlight, not lamplight—something older, alive. It poured through her like breath through glass, bending around a shape she couldn't quite see.

A man stood at its center, his edges blurred by brilliance, his voice little more than a vibration against her ribs.

"Find me," he said.

Or maybe it was, Remember me.

The light brightened until she saw stars spilling from his fingertips—and then everything went white.

When she opened her eyes, morning had already broken.

But the light hadn't left. The scent of lilac hung in the air, soft and persistent, clinging to her skin like memory. Her *bedside candle still smoldered from the night before, sweet and ghostlike, as if it had kept watch while she slept.*

She blinked against the sunlight filtering through her blinds, and for a breath, the faint outlines of constellations burned across her vision.

Lines. Dots. A pattern she half recognized.

Each flicker pulsed with the beat of her heart—alive and synchronized.

She rolled onto her back, one arm over her eyes.

"Not *again,*" *she murmured.* "I *don't even believe in visions.*"

Then it came – a flash.

A corridor made of glass. The sound of rushing water.

A man's hand reaching through light.

Gone as quickly as it came, leaving her breathless and shaken, as though she'd stumbled out of a dream that belonged to someone else.

Her phone chimed before she could think too hard.

SORI: *Morning, star girl. Coffee or caffeine* IV?

ESTELLE: *Both. My brain's fried.*

SORI: Good. *You can fry mine too. Meet in twenty?*

Sori was her favorite contradiction—sharp-tongued and warm-souled, always there, always half-gone. Her honey skin remind Estelle of the barley tea she alwasy drank. They'd met in grad school, two misfits orbiting the same patch of curiosity. Sori was now a systems analyst for AstraCorps—of course she was—and their friendship had evolved into that quiet rhythm where you didn't need to speak every day to feel tethered.

Estelle never told her, but Sori was the closest thing she had to family. Being adopted had left her with roots that never quite took. Her parents loved her in theory, but affection was something learned, not lived. Sori had been the first person who made her feel seen. Sori could somewhat relate being an only child to a strict Korean father and a traveling soulful singing mother. She was the best mix of them, logic and soul.

She'd first met her years ago at a party she hadn't even wanted to attend. Sori's laugh had carried across the room like a small light in all that noise. Estelle doubted she even remembered, but after that night, she began noticing her everywhere—campus hallways, coffee shops, the same seminars. A familiar face in a sea of strangers. So, when she became her roommate in grad school, Estelle didn't question it. She only felt the strange comfort of recognition—like finding an old song you didn't know you knew by heart. Maybe that was why she trusted her so easily. Why she never wondered how Sori always seemed to find her, no matter where she went.

The café was half-empty when she arrived, all brass and plants and soft lo-fi echoing from the speakers. Sori sat at their usual table near the window, laptop open, nails drumming against a half-finished cup.

"You look like you spent the night arguing with the universe," Sori said as Estelle approached.

"I did," Estelle replied, sliding into the seat across from her. "It won."

Sori smirked. "Dreams again? Same man?"

"How do you—?" Estelle said, then hesitated.

Sori raised a brow. "You talk in your sleep when you crash on my couch, remember? Something about 'light,' 'stars,' and 'he's close.' I almost called a priest."

Estelle groaned, hiding behind her coffee. "If I start levitating, you have my permission to record it for science."

Sori grinned, but her tone softened. "You sure you're okay? You've been... different lately. Like you're tuned to a frequency I can't hear."

Estelle looked down at her cup, watching the steam spiral upward like galaxies. "Maybe I am. Maybe I've just been listening too hard."

For a moment, neither spoke. The silence between them felt easy—earned.

Then Sori leaned back and changed the subject.

"Speaking of frequencies, AstraCorps sent an internal alert this morning. Some kind of atmospheric anomaly near the city. Satellite data went weird."

Estelle's hand tightened around her cup. "Weird how?"

"Unstable energy signatures. Pulsing every few seconds, almost rhythmic."

Sori shrugged. "They'll say it's solar interference. They always do."

Estelle glanced out the café window. Above the river, the sky shimmered faintly—like it was breathing.

"Maybe it's something trying to reach us," she whispered.

Sori laughed softly. "Or maybe you just need sleep."

But Estelle's eyes lingered on that faint shimmer long after the laughter faded.

Kael

At forty thousand feet, Kael felt most human.

The prototype jet—codename Astra 02-17—was designed to test atmospheric warp thresholds, but for him, it was just another excuse to chase horizons. The cockpit hummed, soft blue lights blinking in rhythm with his pulse.

Everything in sync. Controlled.

Until it wasn't.

At 43,000 feet, the first glitch hit.

The monitors fuzzed, then cleared, then fuzzed again. His heart skipped a beat—and the jet's telemetry skipped with it.

CONTROL: Reading anomaly on your end, Verran.

KAEL: Yeah, system's catching interference.

CONTROL: External or internal?

KAEL: Working on it.

The pulse came again—steady, rhythmic, almost biological.

He frowned. A heartbeat?

The data spiked, shimmering across his screens.

And then it hit him—something faint, something impossible.

A scent.

Lilac.

It was so brief, so absurd, he almost laughed—but the warmth that followed wasn't mechanical. It was familiar. Comforting.

The static in his headset bent into a low, melodic hum that resonated in his bones.

Not a malfunction. Not interference.

Recognition.

"Come on, baby," he muttered to the jet. "Don't do this to

me now."

The comms hissed, then curled into something like a whisper. A single word—maybe his name, maybe nothing—cut through the noise. And then everything normalized. The readouts leveled, his pulse steadied, and the sky returned to silence.

CONTROL: *All clear on our end, Verran. Whatever that was, it's gone.*

KAEL: *Copy that.*

But he didn't believe it was gone.

He reached toward the cockpit glass, and for a single heartbeat, he saw a shimmer of light that outlined a woman's face—soft and familiar.

He blinked, and it vanished.

Estelle and Kael

That night, neither of them slept.

In two different cities, under the same sky, both stepped outside at the same hour—Estelle barefoot on her balcony, Kael beside the silver wing of his jet.

The air was heavy with static and starlight, as if the world itself were holding its breath.

Without realizing it, they each lifted their hands toward the sky, tracing invisible constellations.

Estelle's fingertips glowed faintly violet.

Kael's reflected gold.

For a moment, the stars brightened—just enough to make them both pause.

Estelle whispered, "Where are you?"

Kael whispered, "There you are."

And though the distance between them stretched across miles and silence, the universe answered in its own language— Two heartbeats, aligned.

Estelle

Back in her lab that evening, the data flickered again.

A signal. A rhythm. Two pulses in perfect sync.

She played the recording through her speakers and felt her own heartbeat match it. Faster, stronger, alive. A low hum rippled through the floor.

Outside, the city lights dimmed.

Above, the same pulse echoed in the clouds, unseen but felt by both of them.

Sori texted her again.

SORI: *Hey. Power just flickered at my place too. Weird coincidence.*

Estelle didn't answer. She stared at the waveform on her screen, one hand over her chest.

It wasn't coincidence. It was connection.

Kael

Back in his hangar, Kael sat on the wing of the Astra 02-17, *watching the faint vibration of residual energy crawl across its surface. He pressed two fingers to the spot just over his heart.*

The same rhythm. The same pulse.

He didn't know who she was or why he felt her across the sky. But somewhere out there, she was awake. And she was listening.

CHAPTER 3
SUBJECT # 9

Estelle

The AstraCorps building didn't look evil. That was the problem.

It gleamed – a tower of brushed metal and mirrored glass, the kind of place people trusted because it looked expensive.

Estelle tightened her scarf against the wind as she crossed the courtyard. The glass doors whooshed open, releasing a wave of sterile air that smelled faintly of citrus and static.

Every time she came here, she told herself she was lucky. Lucky to be contracted. Lucky to touch the stars through data. Lucky to matter.

But lately, luck felt a lot like being watched. She swiped her badge, stepped into the elevator, and watched her reflection split across the mirrored walls — one version of her smiling, one tired, one already looking away.

When the doors opened, Sori was waiting.

"You're early," Sori said, holding two coffees. "Either the universe glitched or you finally learned time management."

Estelle smiled despite herself. "Couldn't sleep. Too much static in my head."

"Static or stars?"

"Both."

Sori's Observation

Sori led her down the corridor toward Lab 6, fingers wrapped around her coffee like a nervous tic. Her eyes flicked toward every camera dome they passed.

Estelle noticed. "You've been twitchy lately."

"Me?" Sori laughed. "You're the one hearing music in light frequencies."

Estelle snorted, nudging her shoulder. "Touché."

They stepped into the lab. The hum was louder here — a faint vibration underfoot, like the building itself was breathing.

Sori pulled up the terminal and motioned for Estelle to look.

On the screen, a familiar figure-eight shimmered, pulsing at intervals.

"It's still active?" Estelle asked.

"Yeah. Voss doesn't know what to make of it. He thinks it's a solar echo."

"And you?"

Sori hesitated. "I think it's calling to something. Or someone."

Their eyes met.

For a heartbeat, the hum matched their pulse.

Estelle looked away first. "You sound like me."

"Guess you're rubbing off," Sori said softly.

Estelle smiled, and something in Sori's chest tightened painfully.

She forced herself to look back at the data. "Dinner tonight? We can actually eat food that doesn't hum."

"Deal," Estelle said.

AstraCorps Central Facility – Observation Chamber 07

The air smelled like ozone and order.

Director Voss stood before a wall that wasn't glass so much as condensed light—polarized to hold, to separate, to make stars obey. Behind it, a pale ribbon of plasma drifted in a sealed column, humming at a frequency the human ear could almost catch if it tried hard enough.

"Do you hear it?" he asked.

Nora Wimbush didn't step closer. "All I hear is containment doing its job, sir."

"It hums," Voss said, not looking at her. "Every time we pull it from the upper atmosphere, it remembers what it was." His reflection fractured across the panel—one face split into a dozen angles. "Memory is a kind of energy. People forget that."

Nora kept her hands folded over a tablet, thumbs still, shoulders squared. She had learned early that stillness looked like respect.

"Extraction Team reports a secondary transient at oh-six-thirty-four. Your jet. The pilot logged interference."

"Verran." A corner of Voss's mouth tightened. "He hears things he isn't paid to hear."

"We can retrain him." Nora *said it like adjusting a dial.*

"Or replace him." Voss *let the hum fill the pause. "Open the* LUMEN *packet."*

She swiped the tablet. The room's side wall resolved into layered telemetry: waveforms braided with map grids; temperature bands; a luminous figure-eight that refused to smooth into noise no matter how aggressive the filter.

"There," she said. "Repeatable pattern. Two spikes, recurs at interval. Then a knot."

"The knot is new," Voss *said softly. "Or newly obvious."*

"Ground echo corresponds to an academic node—Auxiliary Research Lab, City District Four." Nora *zoomed.*

"Contractor access pinged the packet."

"Name."

"Subject 09, civilian alias: Estelle Auryn.*"*

The hum seemed to crescendo. Or maybe Nora *only noticed it now that the name was in the air.*

"Put 09 under closer observation," Voss *said. "Tighten*

permissions. She's been accurate in her noise assessments, which makes her dangerous."

Nora's mouth flicked, almost a smile. "We prefer our contractors observe, not interpret."

"Mm." He watched the column of light twist, as if trying to turn toward something beyond the glass. "And yet interpretation is how human beings survive. That's what makes them...useful."

"Shall I flag TRKRR for contact?" Nora asked, gaze neutral. "Low-grade proximity checks. Social vector first."

"Escalate quietly," Voss said. "If 09 increases her sync, we widen the net. LUMEN feeds GENESIS. That's the order."

Nora didn't flinch at the second word. She only made a note.

"Understood."

As she left, Voss looked at his screen blinking "Project Lumen: Update." A soft pulse flickered across the data feed too bright, too alive for his comfort. Lumen updates rarely meant anything good. He muted the alert with a curt tap, already bracing for whatever problem it would grow into next.

Eetelle

Estelle had come in early, hoping to run another analysis before the building woke fully. But the hum followed her. Every vent, every wall panel seemed to vibrate with invisible static, as if the whole place was whispering.

Her head throbbed. She ducked into the nearest restroom, half for silence, half to escape the eyes of security lenses. She leaned against the cool metal of a stall door and breathed, trying to quiet her pulse.

Through the thin wall, she heard the faint echo of heels – sharp, rhythmic – then the creak of the adjoining stall. A pause. A sigh. Running water.

Then a voice.

"Do this, Ms. Wimbush. Do that, Ms. Wimbush."

It was Nora.

Estelle froze, feet tucked up onto the toilet lid like a child hiding from thunder.

Nora's voice sharpened, mocking.

"Yes, Director Voss. Right away, Director Voss. Anything for the great and powerful god of order."

The sink squealed as water poured, but Nora wasn't washing her hands. She was pacing.

"I worked too damn hard to end up babysitting a bunch of curious little brats who think stars talk back."

The bitterness cracked through her composure. "I should've been running this place years ago. But no—'Wimbush, keep the contractors quiet.' 'Wimbush, manage the handlers.'"

Her voice wavered – tired, small.

"I hate this building. I hate that it hums when I'm alone. Like it knows."

The words hung in the sterile air, almost swallowed by the sound of the hand dryer that followed.

Estelle stayed still, heart hammering. She hadn't meant to eavesdrop, but something in Nora's confession felt raw, almost... sad.

The hand dryer cut off. Silence. Then the click of heels again.

"Get it together, Nora," the woman muttered. "You're not

one of them."

The door swung open and shut.

Estelle waited a full minute before climbing down, her knees shaky. She caught her reflection in the mirror – eyes wide, hair slightly wild.

"She's losing it," she whispered, then immediately hated herself for saying it.

Because underneath the snark was something else.

A thought she couldn't unthink: So am I.

Kael - The Sky Between Worlds

Across the sky, Kael's jet carved through clouds, his radio silent. He'd turned off AstraCorps' *channel hours ago – too many voices, too much interference.*

He banked west, the horizon split into molten gold and violet – the same shades that haunted his dreams.

His sensors picked up a static burst, faint and rhythmic. He glanced at the waveform, and something inside him stirred. The pulse was slow, steady, almost human.

And beneath the static – faint, impossible – he swore he

could smell lilac again.

The scent hit him like turbulence, pulling him backward through time.

A flash, two children running under starlight, laughter rising into the night air.

A girl with dark curls chasing him through a field that shimmered like silver dust.

They tumbled into each other's arms, glowing faintly, their skin haloed in the reflection of a thousand unseen stars.

The memory didn't belong to him. But it felt like his... too warm, too real, as if the stars themselves had loaned it to him for safekeeping. He blinked, the cockpit snapping back into focus. Lilac still lingered, threaded through the recycled air.

He thought of his mother.

She had always smelled like lilacs – soft, constant, like spring after rain.

He'd joined the Air Force to make her proud, to give her something to look up at and say, that's my son, chasing the heavens. But she'd died while he was still overseas, and the sky hadn't looked the same since. Flying was the only time he felt near her – suspended between gravity and heaven, somewhere in the middle of where she might be.

He reached for the photo taped beside his instrument panel – the one of her standing in front of their porch, sunlight in her hair, her hand resting on his shoulder.

The picture sat just beneath his dog tags, swaying slightly with the plane's hum.

"*Is it you,* Mother?" *he whispered.*

The pulse on the monitor answered with silence, but the hum beneath it deepened, almost like a heartbeat returning the question.

SORI

Sori's monitor glowed faint blue in the dark. Sori leaned forward, eyes burning from hours of cross-referencing anomaly packets.

On her screen, a new tag pulsed at the corner of the data feed: SUBJ-09 – LUMEN CLUSTER: ACTIVE SYNC DETECTED

Her stomach twisted. Estelle's name wasn't there, but Sori knew how AstraCorps labeled their contractors – stripped of identity, reduced to numerics.

She whispered the designation under her breath.

"Subject Nine..."

The microphone embedded in her terminal blinked red – recording.

She froze.

She hadn't meant to activate it, but the system was always listening. Every keystroke cataloged, every whisper archived.

[Recording in Progress – Operator ID: TRKRR-04]

She reached for the power toggle and hesitated.

She knew what she was supposed to do: log the report, submit a behavioral summary, note any "deviation events."

She also knew what she wanted to do: erase it.

The cursor blinked like a pulse.

Footsteps echoed in the corridor – light, familiar.

Estelle's voice carried through the half-open door.

"Sori? You down here again?"

Panic sliced through her. She slammed the key sequence to stop the recording, deleting the file as Estelle's *reflection appeared in the glass wall.*

When she turned, her smile was easy – practiced.

"Caught me doing overtime," Sori *said, trying to shut down the monitor. "These readings are ridiculous."*

Estelle *stepped closer, curious. "What readings?" she asked.*

"Nothing interesting. Just system noise."

But the question lingered in Estelle's *eyes as she peaked over* Sori's *shoulder. "Who's Subject Nine? And what's a* TRKRR *report?"*

Sori's *laugh came too quickly. "Internal testing codes and reports. Some of the guys name their dummy variables after comic book characters. You know engineers."*

Estelle *smiled faintly, unconvinced but too tired to press. "You should go home sometime,* Sori*."*

"Yeah," Sori *said, forcing her voice steady. "Yeah, I should."*

When Estelle *left,* Sori *turned back towards the monitor. The deleted file was gone – but the system log wasn't.*

TRKRR-04: *Manual file removal detected.*

Supervisor notification in progress.

Her throat went dry.

"Shit."

She closed the screen, slid her ID card from the reader, and whispered to the empty room,

"She doesn't deserve this."

Then she walked away before the red light could start recording again.

Dinner

The restaurant was small and warm, a mismatched miracle wedged between two high-rises. A candle burned at their table. Sori had chosen the spot, though she wouldn't admit how carefully she'd researched it. She knew Estelle liked quiet corners, dim lights, and food she could pick apart absentmindedly while she thought.

They ate, laughed, and for a while the world shrank to just them — no signals, no labs, no figure-eight blooming across screens. Just two women who found comfort in each other.

"You ever think about leaving?" Estelle asked suddenly.

Sori blinked. "AstraCorps?"

"Everything. The city. The noise. The constant measuring of what's normal."

"Sure," Sori said. "But where would we go? The stars?"

Estelle's laugh was soft, almost sad. "Maybe."

Sori's voice dropped, almost a whisper. "You'd fit there. You glow when you talk about them."

Estelle's cheeks flushed. "Don't be dramatic."

"Not dramatic. Observant."

The candle flickered between them, its scent wrapping the air in warmth.

For a moment, Sori almost reached for her hand – but the red notification on her watch pulsed. Recording active.

Her stomach turned. She'd forgotten to disable the uplink. Every word, every heartbeat, was feeding the system.

She smiled through it, forced and fragile. "You always make the universe sound like it's in love with you."

"Maybe it is," Estelle said with a grin.

"Maybe I am," Sori almost whispered – but swallowed the words with her wine.

Later that night, back at AstraCorps, Sori sat alone in the systems lab. The dinner replayed in her mind – Estelle's laugh, her warmth, her impossible innocence.

The monitor pinged.

SUBJ-09: *Elevated Sync Event – Cross-Frequency Detected*

She froze.

Estelle's heartbeat. The signal. The same pattern she'd just seen in the lab.

"God," she whispered, opening the feed.

The trace connected directly to Kael's jet.

The system tagged it automatically: PAIR MATCH DETECTED.

Sori exhaled sharply, hand trembling over the keyboard.

She could file it. She could erase it.

Either way, it would destroy something she loved.

She whispered to herself, "She deserves to know."

Then she deleted the entry and backed up the file onto her private drive – the first treason she'd ever commit.

Director Voss's Office 22:00

Voss preferred his office unadorned; it made every intrusion feel like an event. The city lay beneath him in a grid of tempered glass and calibrated distance, the sky a thick pane waiting to be acquired.

Nora stood with a crispness that suggested she'd learned to iron while angry. "Perimeter teams report no civilian observation during last night's retrieval," she said. "We'll add a public comms packet to explain power flickers as subgrid maintenance. The press will yawn."

"And the pilot?" Voss asked, removing an invisible piece of lint from his sleeve.

"Verran filed clean. He's competent, if imaginative. We can keep him pointed away from his imagination."

"Hmm." Voss's eyes went unfocused with a familiar arithmetic. "We'll need him for the higher passes. Imagination can be useful when it believes it's logic."

Nora placed the tablet on his desk. "LUMEN yields are increasing. The knot shows up in every pass now."

He turned the tablet and watched the figure-eight bloom, the center brightening second by second as if approaching its own name.

"Patterns," he said. "Everything is a pattern if you wait long enough." He leaned back. "Bring 09 closer. An invitation. A promotion. People are predictable when you give them the respect they think they deserve."

"And if respect fails?"

"Distance fails," Voss said mildly. "Then we shorten it."

Nora's gaze didn't flicker. "TRKR vectors are already in place."

"Of course they are." He smiled without warmth. "Proximity is an art."

"Ethics Board meeting at fourteen hundred," Nora said, as if they had one. "They'll ask what LUMEN is for."

"For?" Voss repeated, as though the preposition offended him. "A tool does not owe a philosophy. It owes performance."

"And GENESIS?" she asked, letting the word be ordinary.

"Insurance," Voss said, and turned to the window. "If the light refuses to be taught, we'll breed it."

Nora's expression stayed exactly where she put it. "Yes,

Director."

"Don't look at me like that, Wimbush," he added without turning. "You came here because you understood what it costs to keep a world from tearing itself apart."

She allowed herself a small, humorless smile. "I came here because I'm good at following orders."

"Same thing."

Voss studied the updated telemetry. The pattern bloomed larger, stronger. The figure-eight now spun faster, its core brightening like a pulse about to ignite.

Nora stood beside him. "Tracker 04 is late with her report."

"She's conflicted," Voss said mildly. "That's useful. Emotion makes her predictable."

"And if it makes her disobedient?"

Voss smiled faintly. "Then she'll lead us right to our target without even realizing it"

He turned to the glass wall, watching the city pulse below like a living circuit.

"Patterns," he murmured. "Always patterns."

Nora nodded in understanding and excused herself.

Estelle and Kael

That night, Estelle dreamed again.

The man of light reached for her – closer now, almost touchable.

"Find me," he said.

Across the sky, Kael woke with the same word on his lips.

And somewhere deep below them, in a lab that hummed like a heartbeat, a file blinked red:

LUMEN SYNC – PHASE TWO INITIATED.

CHAPTER 4
THE FREQUENCY BETWEEN US

Estelle

It felt like the air had been holding its breath.

The sky outside the research complex burned white-blue, the kind of winter light that turned glass into mirrors.

Estelle *tugged her badge against her sweater as the scanner blinked green. Another long day. Another round of pretending she wasn't unraveling.*

"*Morning, sunshine.*"

Sori's voice pulled her back to earth like gravity.

Estelle *smiled despite herself.* "*You're not human.* No *one's this cheerful before eight.*"

"*I'm powered by caffeine and chaos.*" *Sori handed her a steaming mug.* "*And a desperate need for validation from the universe.*"

Estelle laughed—a real one this time—and took the cup. "You're ridiculous."

Sori shrugged, looping an arm through hers. "And yet, here I am. Your emotional support scientist."

They walked the sterile hallway together, passing walls lined with mission plaques and half-dead plants. Sori kept talking, light and animated—about a neighbor's cat that had learned how to open fridge doors, about the new telescope array she wanted to sneak onto—but Estelle's focus kept drifting.

The hum she'd been hearing for weeks was back, faint but insistent, like a song she couldn't quite remember. Every time she tried to ignore it, it would thread itself through the sound of her heartbeat until she couldn't tell which was which.

They reached the glass lab doors. Inside, the rest of the team was already at work, faces washed pale by monitor light.

And at the center of it all stood Nora Wimbush.

She was impossible to miss: tall, angular, with sharp black hair that cut just below her jaw and a smile that never reached her eyes. Even her posture seemed to disapprove of everyone around her. Nora's voice had that syrupy sweetness that only mean people could pull off.

"Miss Auryn," she said without looking up from her tablet. "So glad you decided to join us in the realm of the productive."

Estelle's jaw tightened. "Good morning, Dr. Wimbush."

Nora clicked her pen, once, twice. "I've been informed by Director Voss himself that we're to maintain stricter data compliance moving forward. Which means no extracurricular analysis, no outside consultation, and certainly no... creative interpretations of classified material." Her eyes flicked up. "I trust that won't be an issue."

Sori's expression darkened beside her. "You'd think imagination was a felony in this place," she muttered.

Nora's gaze sharpened. "What was that, Dr. Cho?"

"Nothing, ma'am. Just admiring the décor."

Estelle hid a smile behind her cup.

But the humor drained as soon as Nora turned away.

Because Estelle knew why Voss was suddenly interested again.

The anomalies were back. And she had proof.

That night, when the lab emptied and the hum in her bones grew too loud to ignore, Estelle slipped back inside. The monitors blinked in the dark like tired eyes. She keyed in the restricted code Sori had "accidentally" left open on her tablet and brought up the data stream from the atmospheric probes.

The screen bloomed with motion.

Pulses. Flares. A repeating pattern too deliberate to be natural.

She zoomed in and froze.

The anomaly wasn't just atmospheric. It was being transmitted.

And the source—God, the source—was AstraCorps.

Kael

Two days later, Kael's plane touched down at the same

coastal base. The moment he stepped out, the smell of ozone hit him—the same scent that had filled his cockpit during the anomaly. The same taste of electricity on his tongue.

He wasn't supposed to be here. This wasn't his sector anymore. But when Voss called him in for a "routine debrief," he didn't argue. Curiosity was a weakness, and Kael had always been weak for mysteries.

The lobby was sleek, sterile, empty except for a receptionist typing too fast and an ID badge scanner that blinked red until he swiped again. "Welcome back, Captain Verran," said a voice overhead. "Director Voss will see you shortly."

He never got that far.

Because when the elevator doors opened—she was there.

Estelle

She almost didn't look up.

Her hands were full of files she shouldn't have been reading, and her mind was still spiraling through the realization that AstraCorps had been fabricating their own data streams to cover the anomalies they'd caused. She was halfway through the corridor when someone brushed past her shoulder.

Her breath caught.

It wasn't a touch—it was a current.

She turned, heart hammering, and locked eyes with him.

The world fell silent.

It wasn't recognition, not in the ordinary sense. It was something deeper—like remembering a song you'd loved in another lifetime. The air between them pulsed once, then again, matching the rhythm she'd been hearing for weeks.

He looked at her the same way she looked at the sky when she needed to breathe—like he'd been waiting to find her there.

For a moment, neither of them moved.

And then, as if the universe itself exhaled, the lights above them flickered.

"Sorry," she managed, her voice a whisper. "I didn't see you."

Kael blinked, as if waking from a trance. "No, that was—my fault. I should've... watched where I was going."

His voice. It vibrated in her chest the way her dreams did—low, familiar, wrong and right all at once. She tried to step back, but her hand brushed his sleeve, and the world shifted.

A flash—stars, a voice, the same light from her dream.

Then darkness.

She gasped, staggering back. He caught her before she could fall.

"Are you okay?" His touch grounded her. "You look like you've seen a ghost."

"Maybe I have," she said softly.

Their eyes met again, and this time she saw it—the faint shimmer of light threading through the air between them.

Neither spoke. Neither needed to.

Deja vu was too small a word for it.

Sori

From across the hall, Sori saw it happen.

She was halfway through pretending to read a tablet when

she noticed the way Estelle's shoulders froze, the way her expression softened into something Sori had never seen before—equal parts awe and ache.

And the man—light-skinned, tall, dreads catching the fluorescent light—looked like he'd just found the missing half of a map he hadn't known he was carrying.

Sori's chest tightened. She'd teased Estelle for being a dreamer, for chasing cosmic ghosts, but this—this wasn't just chemistry. This was gravity.

She cleared her throat loudly, breaking the moment. "Hey, star girl," she said, voice teasing. "You gonna introduce me or just keep staring like that's a crime scene?"

Estelle blinked, flustered. "Oh—uh, Sori, this is—"

"MyKael Verran," he said, extending a hand. "Test pilot. I think I'm lost."

Sori shook his hand, smirking. "If you're lost, you found the right person. Estelle here specializes in cosmic directions."

Estelle shot her a look that promised revenge. "Sori."

But Sori's grin softened. Beneath the sarcasm, her eyes were kind. Protective. She could already sense whatever this was—it wasn't fleeting. And she wasn't sure whether that thrilled her or terrified her for her friend.

Ms. Wimbush

By the time they reached the lab, Nora was waiting.

"Miss Auryn," she snapped, voice slicing through the air. "A word."

Sori whispered, "Dead woman walking," under her breath and slipped out.

Estelle straightened. "Yes, Dr. Wimbush?"

Nora crossed her arms. "Director Voss has concerns about your recent curiosity. We've noticed unauthorized log-ins under your ID. Care to explain?"

Estelle's throat tightened. "I was verifying an anomaly."

"You were snooping." Nora's smile was acid-sweet. "You see, Miss Auryn, AstraCorps has invested considerable time in maintaining the stability of our research. The last thing we need is a curious little stargazer undoing it with... idealism."

Estelle met her eyes, calm but trembling inside. "You mean truth."

Nora leaned closer, voice dropping. "You have no idea what truth costs."

For a flicker of a moment, something unguarded crossed her face—fear, maybe—but it vanished behind that cold veneer. "You'll stay in your lane, Miss Auryn. Director's orders."

As Nora left, the hum returned, louder than ever. The lights overhead flickered again.

Estelle turned toward the window and saw the horizon shimmer—the same pulse Kael's instruments had captured. The same pulse that was calling her name.

Kael

Later that night, standing outside the facility, Kael looked up at the sky.

He didn't understand what had just happened in that hallway, but he knew one thing with absolute certainty. He'd seen her before.

Not in this life. But in another one.

And whatever bound them—whatever AstraCorps was hiding—wasn't finished with them yet.

Estelle

By the time Estelle finally made it home, the night had stretched into that soft hour between exhaustion and dream. The city outside hummed faintly, lights flickering in a sleepy rhythm.

She sat on the couch, sketchbook open across her lap. She didn't plan to draw him – not really. But every time she tried to trace something else, her hand betrayed her.

Kael's face took shape in graphite: the sharp line of his jaw, the faint tilt of his mouth when he almost smiled, the quiet in his eyes that felt older than he looked. She shaded in the details slowly, carefully, until the paper almost seemed to breathe.

When she finally dozed off, the sketch was still beneath her hand – a half-finished portrait of the man she'd known for two days but somehow remembered for years.

The front door clicked open.

Sori's voice floated in, half-yawn, half-groan. "You're lucky I bring good coffee, woman."

She dropped her bag on the counter and looked around – the scattered papers, the faint scent of pencil and lilac.

Then her eyes caught the sketch.

"Oh, Estelle..." she whispered, her expression tightening.

She didn't mean to frown – but she did.

By the time Estelle stirred awake, Sori was in the bathroom, water running. Estelle stretched, blinking groggily.

"Sori?"

"Your spare key still works," Sori called over the sound of the shower. "Creepy much?"

Estelle laughed, heading toward the sink. "You breaking and entering again?"

"Only when I'm concerned," Sori said lightly. "And maybe a little jealous. That's the guy?"

Estelle hesitated, glancing at the sketch still open on the coffee table.

"Yeah" she said softly. "There's just something about him. .

Like I've seen him before the other day."

The water cut off. A pause followed – longer than it should have been.

When Sori stepped out, towel around her shoulders, her expression was carefully composed. But her eyes... they told another story.

"Be careful with that kind of feeling," she said, voice steady but tight. "Sometimes the things that feel familiar aren't meant to be remembered."

Estelle looked at her, puzzled. "What's that supposed to mean?"

Sori forced a smile that didn't reach her eyes. "It means maybe get some sleep before you start drawing strangers in your dreams."

She turned away, but Estelle noticed the flicker in her expression – the way her gaze lingered on the sketch one last time, not with curiosity... but with recognition

CHAPTER 5
GRAVITATIONAL PULL

Kael

The day after their first encounter, Kael found himself standing outside the small campus café, pretending to scroll through his phone while hoping she'd appear.

And then she walked in.

Estelle.

Her blackish-brown curls were pulled into a loose bun, soft tendrils escaping to frame her face in the late morning light. Her skin caught the sun like it remembered warmth, smooth and radiant, and her oversized sweater swallowed her hands as she gestured animatedly at the barista about the "moral crime" of burnt coffee.

Kael couldn't help but smile. He stayed back, not wanting her to think he was following her like some creep who couldn't take a hint. From his quiet corner beneath the awning, he just watched her – the way she laughed with her whole face, unguarded, real.

She smelled faintly of cocoa and caffeine – like late nights and comfort. It fit her somehow.

He caught himself smiling wider, shaking his head.

Yeah, he was gone.

He made a silent promise then:

Next time, he wouldn't just watch from the shadows.

He'd be the one buying her coffee.

Estelle

Three days had passed since she'd first seen him in that hallway, yet she couldn't stop replaying the moment.

His eyes.

The pulse that synced with hers.

The quiet certainty that she'd known him before, maybe in another time.

She'd buried herself in work, pretending normalcy, pretending she didn't feel the universe humming under her skin. Even Sori had noticed.

"You're jumpy," Sori said one afternoon, balancing a cup of tea on her knee. "You flinch every time the lights flicker. You sure you're not possessed?"

Estelle shot her a look. "Funny."

"Just saying." Sori sipped. "If you start floating, I'm charging admission."

Estelle smirked despite herself. "You'd really sell tickets?"

"For science." She said smilng warmly.

Their laughter softened the tension, and for a while, it almost felt ordinary again. But that night, walking home through the narrow back street that cut behind the labs, the hum returned – stronger, closer, like it was following her heartbeat.

She paused beneath a streetlamp, her breath misting in the cool air. "Okay," she whispered to no one. "Either I'm losing my mind, or something's—"

"Following you?"

The voice came from behind her. Smooth, deep, familiar.

She spun around. He was leaning against the rail, half in shadow, hands shoved into his jacket pockets like he'd been *standing there awhile. The same man.*

Kael.

Her chest tightened before she could think. "Wow," she said, folding her arms. "I was just talking about stalkers. Nice of one to show up on cue."

Kael's mouth curved into a slow grin. "That's fair. But in my defense, I wasn't stalking. I was... tracking."

"Tracking?" She arched an eyebrow. "Oh, that sounds so much better."

"I know how it sounds." He rubbed the back of his neck, eyes warm despite his nerves. "I just... I couldn't shake it. You."

"Me?"

"I know this'll sound insane, but..." He looked up, the streetlight glinting off his lashes. "I've seen your face before. In dreams. Before that day in the hallway, before I even knew you existed."

Her teasing expression faltered. "Dreams?"

"Yeah." His voice was quieter now. "Every time I flew, there was this light. Same as the one that burned through my

cockpit the day the systems glitched. And in the middle of it – you. Your face."

For a moment, she couldn't breathe. She'd spent weeks waking from the same dream – a man surrounded by light, calling her name.

He stepped closer. "I told myself I was just tired, that I imagined it. But when I saw you standing there in the lab... I knew I hadn't."

Estelle forced a smile, her heart racing. "That's... definitely crazy."

"Yeah," Kael admitted softly. "But the thing is – I think you've seen me too."

Her eyes widened. He wasn't guessing. He knew.

"How did you–"

"Because the first time I looked at you, it didn't feel new," he said. "It felt like remembering."

They stood there, silent except for the noise of the city and the faint buzz of the streetlamp above. The world around them seemed to hold still.

Estelle broke the silence first. "Well," she said, voice low. "If you're lying, you're good at it."

He smiled. "I wish I were."

"I was thinking," he began, then stopped, his voice rougher than he intended. "That maybe we should... you know, talk somewhere that isn't surrounded by computers and government secrets."

Estelle blinked, caught off guard by the sudden vulnerability in his tone. "Are you asking me out, Captain Verran?"

He smiled – the kind that started at one corner of his mouth and didn't quite reach his eyes. "Only if you're saying yes, Professor Auryn."

She tilted her head, pretending to weigh the offer, but the warmth in her chest betrayed her. "When and where?"

Kael exhaled like he'd been holding the question for days. "There's a small café near the observatory. Quiet. They open the rooftop at night – good coffee, better view."

She studied him for a heartbeat too long and realized – he'd already planned it. The place, the time, maybe even the weather.

"You've been thinking about this," she teased.

He didn't deny it. "Maybe."

And in that small admission, Estelle felt something shift. The man who'd fallen through her dreams and the pilot who'd flown through her signal – he was real, standing in front of her, fumbling over a date.

"Okay," she said, the corner of her mouth lifting. "Coffee sounds perfect."

He grinned, a rare and unguarded thing. "Tomorrow night, then?"

"Tomorrow night."

As he left, the door closed with a soft click. Estelle pressed her hand over her chest, feeling her pulse flutter beneath her palm – not cosmic, not celestial, just the thrilling, terrifying ache of something beautifully human.

The Rooftop

The café rooftop was smaller than Estelle expected – a scatter of wrought-iron tables, warm string lights, and the soft music of the city below. The kind of place that made the world feel manageable, like life could pause long enough for two people to breathe.

Kael was already there, pretending not to check the time for the third time. He told himself he was calm – until he saw her.

Estelle stepped through the rooftop doors like a vision the night had been saving. Her blackish-brown curls framed her face in loose waves, and the city lights kissed her skin like gold. She wore a fitted midnight dress that hugged her in all the right places, a cropped cream sweater draped casually over her shoulders. Effortless. Radiant.

Kael forgot how to breathe for half a second.

She carried herself like the world turned just a little slower when she walked into view.

"Hey," she said, smiling. "You clean up nice."

"Hey yourself." He tried to deepen his voice, adding a touch of that deliberate, masculine gravity – but it cracked slightly on the last word.

Estelle's brow arched, then she laughed, warm and musical. "Was that your grown man voice?"

He chuckled, rubbing the back of his neck. "Maybe."

"Well, it's adorable," she teased. "You sound like you're trying to sell me a used car."

"Good thing I'm not," he said, recovering quickly. "Though I can promise the ride's worth it."

She laughed again, covering her mouth – but the blush in her cheeks betrayed her. There was something about him, something easy. He didn't try too hard, didn't fill silence with noise. He made her feel seen.

The waiter came and went, bringing coffee and something sweet neither of them touched. They were too busy stealing glances. Every time their eyes met, something fluttered – a spark, a pulse, the kind of nervous warmth that had no explanation.

At one point, the city grew loud – traffic, laughter, clinking glass – but between them, everything went quiet. The world dimmed. It was just Kael and Estelle, smiling like two people who'd just remembered what joy felt like.

Her gaze wandered over him – the sharp lines of his jaw, the faint dimple in his cheek when he grinned, the warmth in his light brown eyes.

He watched her right back – the way her lips curved when she was holding back a laugh, the small gap in her smile that made it even more perfect, the deep, soulful brown of her eyes that could stop his pulse.

He was grinning so hard it almost hurt. He shook his head, trying to play it cool.

"I enjoy being the object of your gaze," Estelle said, teasing, *her voice laced with laughter.*

Kael leaned back, mirroring her smirk. "Can you blame me? I'm just trying to keep up."

And then they laughed again – not nervous laughter this time, but something real. The kind that comes from the unspoken knowledge that they were both gone already.

Months slipped by in the easy rhythm of shared mornings and long nights. What started as coffee dates turned into constellations of moments stitched together – movie marathons that ran until sunrise, quiet dinners where they forgot to eat because they were too busy laughing, and late-night walks through the city where their hands found *each other without thinking.*

Kael learned that Estelle couldn't cook to save her life but could quote entire sections of astrophysics journals from memory. Estelle discovered that Kael hummed under his breath when he focused – a low, tuneless sound that somehow made her feel safe.

Their favorite night was the space exhibition downtown. Estelle dragged him there insisting it was "for research," but Kael caught the sparkle in her eye when they stepped beneath the planetarium dome. He whispered something ridiculous about naming a star after her; she called him a flirt and laughed until she snorted, which only made him fall harder.

Outside that dome, though, the sky was changing. The frequency of celestial anomalies – lights, distortions, energy pulses – had begun to climb. Reports surfaced across networks and newsfeeds, scientists whispering about patterns that didn't fit any known model. Even Sori mentioned it once in passing, showing Estelle a report over lunch.

But Estelle barely looked up from her coffee. She smiled, brushed it off, and said, "Maybe the universe is just showing off."

The truth was, she didn't care to chase the unknown anymore. Not when something real – something warm – was finally within reach.

In time, their lives folded into one another seamlessly – her sketches finding space on his desk, his jacket always hanging by her door. They had their own kind of gravity now, pulling them back together no matter how far apart their days began.

It wasn't perfect – love never was – but it was theirs. And for a while, that was enough.

Sori

Sori *wasn't stupid.*

She'd seen the way Estelle *had been smiling at her tablet lately, the way her eyes went unfocused in the middle of a conversation like she was hearing someone else's voice.*

And when she walked into the lab on Friday *morning and*

caught Kael leaning over Estelle's monitor, the two of them standing closer than workplace regulations would ever allow, she almost dropped her coffee.

"Wow," she said loudly, "is this the part where we all pretend not to notice the cosmic tension in the room?"

Estelle's cheeks flushed pink. "Sori—"

Kael straightened. "We were just talking."

Sori raised an eyebrow. "Sure. That's what people call it these days."

Estelle glared, but Sori only grinned wider. She liked him instantly. Not because of the way he looked at Estelle – like he'd been waiting for her – but because he looked human. Grounded. Maybe even good for her friend.

"Relax," Sori said, softening. "You think I'm gonna snitch to Nora? Please. I'd pay to watch that woman trip over her own ego."

"Speaking of Nora..." Estelle murmured, eyes darting to the hallway. "She's been asking questions. About the data stream I was analyzing."

Sori's grin faded. "Yeah, I know. Word is, Voss put her on a leash. He's paranoid about someone 'discovering sensitive information.'"

"Sensitive like what?" Kael asked.

Sori hesitated, then leaned in. "You didn't hear this from me, but... rumor is AstraCorps has been running energy experiments using atmospheric anomalies – the same ones you two keep talking about."

Estelle frowned. "Manipulating them?"

"More like trying to harvest them."

Kael's jaw tightened. "They're pulling energy from celestial events?"

Sori nodded. "Something about controlled containment. They call it 'Project Lumen.' I thought it was theoretical – but if those anomalies match your flight patterns..."

Estelle sat down hard, the pieces falling into place. "They're not studying the anomalies," she whispered. "They're causing them."

Sori met her eyes. "And maybe that's why they're so scared of what you found."

Kael

That night, Kael met Estelle on the rooftop above her apartment building. The air was sharp with sea breeze, and the stars blinked faintly through thin clouds.

She looked different up here – more at ease, less scientist, more soul.

"They're harvesting starlight," she said, voice shaking with anger. "Celestial radiation, gravitational currents – trying to recreate a Lumen Field. That's why the sky pulses whenever the readings spike. They're pulling energy from living stars."

Kael exhaled slowly. "That explains the interference during my flight. The plane wasn't glitching – it was crossing through their field."

"And you were the catalyst."

He looked at her sharply. "Me?"

"You and I were both connected to it. Every time you were in the air, the anomalies spiked. Every time I dreamed, they flared."

She stepped closer. "Whatever AstraCorps is doing, we're a part of it."

Kael studied her face, the glow of the city lights reflecting in her eyes. "You think that's a coincidence?"

"No." She shook her head. "I think it's fate."

A silence fell between them – electric, weightless.

Kael took a step forward, the pull between them growing louder, filling the space like invisible music.

"You know," he said softly, "for someone who once accused me of being a stalker, you're doing a terrible job staying away."

She laughed under her breath. "Yeah, well... maybe I'm drawn to you too."

He smiled. "Guess we're both crazy then."

"Guess so."

For a moment, everything faded – the city, the secrets, the danger. Just two souls standing under a sky that had seen them before. But beneath their feet, deep below the facility, machines whirred and lights pulsed in rhythm with their hearts – proof that AstraCorps' greatest experiment was no longer contained.

CHAPTER 6
THE KEEPER

They would say later that it was curiosity that led them there. That Estelle Auryn's questions were the first cracks in the glass.

But it was never just curiosity.

It was hunger. A quiet, endless ache to know.

She had always been drawn to the unknown – the stars, the data streams, the patterns hidden in static. To her, every anomaly was a whisper, every unanswered question a pulse begging to be felt. She used to tell Sori that science wasn't about discovery, it was about remembrance – about finding what the universe already knew and had simply hidden from impatient eyes.

So when the anomalies began—when the satellites flickered and the sky itself seemed to hum—Estelle didn't run from it. She leaned closer. She listened.

And what she heard changed everything.

AstraCorps wanted to call it interference, radiation, coincidence.

But she saw something deliberate in the noise.

She saw rhythm. Structure.

Purpose.

Curiosity was supposed to be harmless.

But hers became the match that found dry air.

That was how they ended up there—three souls huddled in the glow of forbidden monitors, breaking into the most guarded system on Earth. Not for greed. Not for glory. But because something in Estelle refused to stop asking.

And because sometimes, the only way to find truth… is to tear down the walls that keep it caged.

Sori

If anyone had told Sori she'd be breaking into AstraCorps' Level Four archives for a friend, she would've laughed.

But Estelle wasn't just anyone.

They'd shared too many late nights, too many whispered

theories about what the stars were hiding. Estelle's curiosity burned too brightly to smother. Sori had always admired that—envied it even.

So when Estelle came to her, eyes blazing and voice shaking with the need to know, Sori didn't hesitate.

"Alright, star girl," she said, sliding a flash drive across the desk. "This one gets past firewall two. I'm not saying where I got it, and if you ask, I'll deny everything."

Estelle blinked. "You'd really risk your clearance for this?"

Sori grinned.

"Please. I've been dying to see what the mothership's hiding."

Kael, standing by the door with his arms crossed, gave her a skeptical look. "And you're sure this won't trip any alarms?"

"Not unless you two start live-streaming it." Sori rolled her eyes. "Now hush and let the genius work."

Her fingers danced over the keyboard, bypassing encryption walls that would've made most engineers cry. When the screen flickered open, the header glowed red:

PROJECT LUMEN – INTERNAL ARCHIVES

Sori leaned back with a whistle. "There she is."

Estelle's breath caught. "My God..."

The files weren't research logs. They were histories—names, dates, ages. Photographs of children under sterile lights, their irises glowing faintly like captured stars.

Kael leaned over Estelle's shoulder. "What am I looking at?"

Sori's tone softened. "Experiments. Early-stage genetic testing. These aren't adult volunteers—they're kids."

Estelle scrolled, her hand trembling. "Lumen-Children Initiative... exposure to concentrated starlight energy... radiation mapping..." She swallowed hard. "They altered their DNA."

Kael muttered, "They were trying to make something. Or someone."

Sori's voice dropped. "Half of these files are redacted. But look here—each subject's record ends the same way: Relocated for long-term observation."

Estelle scrolled faster. Coordinates blinked near the bottom of one file. "These aren't labs."

"They're locations," Kael said. "Homes."

Sori frowned. "So they tracked them. Probably to see if the experiments lasted."

Estelle's eyes darted to a line buried at the end of the document: Associated Personnel: The Keeper — *Defector. Last Known:* Outpost 14, West *Ridge.*

"The Keeper," *she whispered.*

Kael glanced at her. "You know him?"

"No," she said. "But if he defected, he might've taken something with him. Proof."

Sori shook her head, incredulous. "You're talking about chasing a ghost across state lines based on coordinates from a fourty-year-old file?"

Estelle's *jaw set.* "Yes."

Kael's lips curved faintly. "Then I guess I'm driving."

Sori groaned, rubbing her temples. "Of course you are." But the humor faded from her eyes. "If you're serious, I'll cover you. I can spoof your ID *logins so it looks like you're still* on-site. *But I must remain behind to do so."*

"Are *you sure?"* Estelle *asked, overwhelmed.* No *one had* ever risked their entire career—let alone their freedom—for her before.

"Star girl," Sori said, smiling softly, "I've been covering your impulsive butt since grad school. What's one more felony?"

Estelle laughed, the sound fragile but real. "You're impossible."

"Admit it," Sori said, bumping her shoulder. "You'd be lost without me."

Estelle

By morning, her mind was made up. She packed light—portable scanner, data drive, and a map printout of West Ridge. Kael met her at the loading bay, dressed in civilian clothes and the quiet determination that always clung to him.

"You sure about this?" he asked.

"No," she said honestly. "But I need answers."

"Then we'll get them."

Sori appeared at the doorway, tablet in hand.

"I rigged your access for forty-eight hours. After that, Voss's system will notice the gap. I'll handle Nora if she asks.

"Sori..." Estelle said as tears formed in her eyes.

"Don't get emotional Estelle, you'd do the same. Just—be careful, okay? And text me when you can."

Estelle hesitated, then stepped forward and pulled her into a hug. "You're my favorite person on this planet."

Sori hugged her back, hiding the tremor in her hands. "Try to keep it that way."

As Estelle turned away, Sori called after her, "And if this Keeper guy turns out to be an axe-murderer, I'm deleting our group chat and denying everything!"

Estelle laughed, waving over her shoulder. "Deal!"

The laughter echoed down the hall long after she'd gone.

They left the city before sunrise. The highway cut through mountain ranges that glowed with frost, the world around them waking in slow silver light.

Estelle sat in the passenger seat, scrolling through the decrypted files again. "Look at this. Some of these Lumen-

Children subjects have date stamps decades apart. They weren't abandoned—they were monitored." Kael tightened *his grip on the wheel. "By who?"*

She turned the screen toward him. "The Keeper's name keeps popping up on reports about 'containment protocols.' Maybe he was the one who helped them hide... or made sure no one escaped."

"Or both."

As time passed, Kael flicked through stations until an old, grainy love song slipped through the speakers—soft, aching, familiar in a way that made no sense.

Estelle frowned. "I've never heard this."

"Me neither," Kael murmured.

Yet somehow, both of them began humming along—perfectly in sync.

They stopped at the same moment, startled.

Their eyes met, warm and searching in the dim cabin light.

When Estelle's fingers brushed his on the console, a spark jumped—gentle, electric.

The radio snapped. Static flared, lights flickered, the song collapsing into broken noise before dying altogether.

Silence settled between them—heavy, intimate, charged.

Neither spoke.

Neither moved.

But the air between them felt like the echo of a kiss that hadn't happened yet.

They drove in silence for a while, the rhythmic hum of the tires matching the faint pull in their chests.

Estelle looked out at the rising sun. "You ever get the feeling you've been here before? Like... the roads, the light—it all remembers you?"

Kael smiled faintly. "Every day."

She turned toward him. "Do you think that's why we found each other? Because the universe wanted us to remember?"

He met her gaze for a heartbeat that stretched too long. "Whatever it wants us to remember, I think it never should've been forgotten."

Arrival

By dusk, the highway narrowed into a gravel path that

wound through pines heavy with mist. Their headlights cut through the fog until the road ended at a rusted gate overgrown with ivy. Beyond it stood a cabin half-hidden in the trees, its windows dark.

"This is it," Estelle said, her pulse quickening.

Kael killed the engine. "Looks abandoned."

They climbed out, the air crisp with the scent of rain and pine. Somewhere in the woods, a low hum vibrated beneath the quiet—a sound that felt familiar in their bones.

Estelle stepped forward, her hand brushing the gate.

"The Keeper lives here."

Kael glanced at the cabin. "Let's find out what he's keeping."

A single light flickered to life inside, soft and gold. The forest was silent except for the hiss of wind through the trees. Kael and Estelle approached the cabin slowly, flashlights cutting pale lines through the dark.

Kael rapped his knuckles against the door. Once. Twice. On the third knock, it creaked open. A faint smell of dust and cedar drifted out, mingling with something sharper—metal, ozone, age.

Kael glanced at her. "Stay behind me."

They stepped inside. The floorboards groaned underfoot.

The cabin looked abandoned at first glance—shelves lined with old journals, half-burned candles, and equipment that hummed faintly, still alive.

Estelle's light swept across a wall covered in pinned photographs and maps.

She froze. "Kael... look at this."

The wall was a mosaic of faces, charts, and handwritten notes—threads of red string connecting names and coordinates. Dozens of them. Children. Teenagers. Adults. Each photo was stamped with a word: SUBJECT. And below each one, a symbol—the same spiral pattern that appeared in the Lumen files.

"Oh my God," Estelle whispered, her voice trembling. "He was tracking them. The Lumen Children... the ones from the files. It's all real."

Kael stepped closer, his jaw tightening. "This wasn't just research. This was surveillance."

Before she could answer, something moved behind them.

A blur. A whisper of breath.

Kael barely had time to turn before a figure lunged from the shadows, slamming into him and driving them both into the wall.

"Kael!" Estelle cried out.

The man's grip was like iron, arms wrapped around Kael's throat. But Kael's instincts took over—years of muscle memory, reflex honed by chaos.

He slammed his elbow into the attacker's ribs once, twice, twisting free and grabbing the man's arm. The movement was fast, clean, lethal. In seconds, the intruder was on the ground, pinned face-down, Kael's knee in his back.

"Stop!" Estelle said, her voice shaking.

Kael's breathing was hard, steady. "We're not here to hurt you," he said through clenched teeth. "We want answers."

The man coughed, struggling under his weight. "Answers..." he rasped. His voice was old, cracked by years of smoke and silence. "You came looking for the wrong kind."

Kael pressed down just enough to keep him still. "You're

the Keeper, aren't you?"

The man gave a pained laugh. "That's what they used to call me."

Estelle stepped closer, her heart racing. His face came into view under the dim light—lined with age, streaked with dirt, eyes pale gray like weathered steel.

The Keeper looked up at her and stilled. "You... you shouldn't exist."

"What?"

His gaze flickered between her and Kael, disbelief written deep in the wrinkles around his eyes. He raised a trembling hand, patting Kael's arm in a gesture of submission.

"I'll tell you everything. Just let me breathe."

Kael hesitated, then eased his grip. The Keeper rolled onto his back, gasping, rubbing at his throat. The Keeper's hands trembled as he gripped the edge of the table. "I've spent my life trying to make sense of what he did—of what they made. The world thinks AstraCorps is mapping the stars. They're not. They're trying to rebuild them."

Estelle's eyes glistened. "And the anomalies?"

"They're ripples. Fractures in the barrier between what we are and what we were meant to be. Every time you feel that hum in your bones, every time the sky pulses—it's not interference. It's the Starborn calling each other home."

He turned toward her, voice softening. "And somehow, child, I think you've already begun to answer."

Estelle couldn't speak. The room seemed to shrink around her, air thick with truth and fear.

Kael moved closer, his hand brushing hers. "If what you're saying is true," he said to the Keeper, "then AstraCorps isn't just studying anomalies. They're playing as gods."

The Keeper nodded once, eyes shadowed. "And gods," he whispered, "are the one thing mankind can never control."

For a long moment, he just stared at them—two strangers who weren't strangers at all. Then he spoke.

"You think AstraCorps made the anomalies. You're wrong." His voice was rough but steady. "They amplified them. Fed off them. But it started long before AstraCorps existed. Long before you were born."

He stood, slow and deliberate, moving to the table cluttered with papers and a cracked glass globe that shimmered faintly with starlight.

"Nearly seventy years ago," he said, "they found something that changed everything. Two children. Twelve years old. Boy and girl. They weren't normal—not by any human measure."

Estelle's breath caught. "What do you mean?"

"They were Starborn." The word came out reverent, trembling. "Carriers of stellar fragments. Living conduits between matter and energy. They glowed when they dreamed. When they touched, instruments broke. When they cried, entire systems lost power."

He turned to face them, eyes distant. "AstraCorps captured them. They said it was for study. But it was torture. They wanted to map the frequency of creation itself—the code of the cosmos written in human DNA. They wanted to weaponize it."

Kael's jaw clenched. "And they used their DNA to create more subjects."

The Keeper nodded slowly. "They took samples. Tried to

replicate the power. Spliced starlight into newborns. Children, soldiers, anyone they could control. Most of them died screaming. The ones who survived... changed."

Estelle felt sick. "The Lumen Project."

He met her gaze, his expression unreadable. "The experiments you found—those were built on the blood of those two children. Their names were erased. Their existence buried. But their legacy... lived on. The fragments still call to each other, even now."

Kael took a step forward. "And you tracked them. The others."

"I tried to save what was left of them," the Keeper said quietly. "To find the ones who survived. To hide them before AstraCorps could harvest more."

Estelle's voice wavered. "Why hide this? Why live like this?"

He turned away, his shoulders sagging.

"Because my father was the one who killed them," the Keeper said, his voice low and steady despite the fracture beneath it. He exhaled slowly. "And then they made me one of the Lumenborn... to see if I could locate the rest. I was *only 6 years old. The merge was sucessful—Starborn DNA fused with human," he said quietly. "The scientists labeled us Lumenborn to match their project name. But we... we call ourselves Astraborn. I failed to locate the other Starborn. Fabricated Starborn... we don't have their abilities. Not even a quantified fraction."*

The silence that followed was thick enough to drown in.

"My father," he continued, voice breaking, "was a commander for the early division that became AstraCorps. When the children resisted, when their power began to merge—he was ordered to 'contain' the experiment. He pulled the trigger. And the light went out."

He turned toward the window then, his voice lowering to a tremor. "But this story... this story only starts there."

He looked away, shame pulling at his expression like gravity. The cabin had grown cold, the air still and electric. Shadows from the single lantern danced across the walls, warping the faces of the children in the photographs.

The Keeper stood by the window, his hand pressed against the frost-glazed glass. When he spoke again, his voice was lower, heavier.

"The story didn't end with the children my father killed," he said. "There was... another occurrence."

Estelle and Kael exchanged a glance.

"Thirty-five years ago," the Keeper continued, "two subjects appeared—older this time. Eighteen, maybe nineteen. A boy and a girl. They'd been adopted separately, lived states apart, but every few nights both would disappear for hours—drawn to the same coordinates, as if by magnetic force."

He turned toward them, his eyes pale and unreadable.

"They didn't know each other, but their hearts remembered what their minds had forgotten."

Estelle's voice was soft. "You mean—"

"I mean," the Keeper said, "they were reborn. Same pattern. Same pulse. Same fusion of light. AstraCorps found them, too. It's what they do—they follow the echoes. But this time the pair didn't survive the capture. Their bond was too strong. When they were separated, their bodies... imploded. Almost like an act of suicide. Matter collapsing inward, as

if the universe refused to let them exist apart."

Estelle's breath hitched.

Kael clenched his fists.

The Keeper's gaze shifted between them. "You feel it, don't you? That gravity. That pull that never makes sense."

Neither spoke, but the silence answered for them.

He moved to a table cluttered with old schematics and fragments of glass that glowed faintly, pulsing in rhythm with the hum in the air.

"This is what AstraCorps never told the public," he said. "Those children—the twelve-year-olds from the first incident—were not just anomalies. They were messengers. Their stellar DNA allowed AstraCorps to ping the coordinates of their origin world. The planet they came from was called Lyratheon."

Estelle whispered the name like a prayer. "Lyratheon..."

The Keeper nodded. "A world made of light and resonance. A living star that birthed its own children. But when AstraCorps found it, they panicked. Their scientists claimed that Lyratheon's orbit had shifted, that within centuries, it would absorb Earth's energy field to sustain its own. Humanity did what humanity always does when faced with something greater than itself—"

"They destroyed it," Kael said quietly.

The Keeper's eyes glistened. "They launched a stellar strike. Broke Lyratheon into a thousand pieces. Shattered the very star that gave you life."

Estelle pressed a hand to her chest. "The anomalies... the fragments in orbit..."

"Exactly." The Keeper's voice grew more fervent. "The debris entered Earth's atmosphere over decades, invisible, scattered. Those fragments didn't die. They became. They embedded themselves into the planet's magnetic field, into human cells. Into bloodlines."

He looked at the two of them, his gaze sharp, reverent, almost afraid.

"My theory," he said, "is that Lyratheon wants to be reborn. But it cannot do so alone. It calls through its fragments—through you. Two lovers destined to find each other again and again, drawn across lifetimes and galaxies to reignite *the core that was lost. The star within."*

Estelle swallowed hard. "You're saying..."

"That you are those fragments," the Keeper finished. "Pieces of Lyratheon given form. Your connection is not coincidence—it's cosmic design. You've always been different, haven't you? Always better together."

Kael took a step closer to Estelle, his voice barely above a whisper. "You mean we're not just connected. We are the connection."

The Keeper nodded. "Two halves of one celestial being. You will always find each other, no matter the body, the time, or the world. Because the universe itself needs you to. The Lyratheon cycle will continue—life after life—until you complete what was begun."

Estelle's heart thudded in her chest. "And what is that?"

"To restore balance," the Keeper said. "To rebuild what was destroyed. Lyratheon's energy can be reborn only when its original fragments—the two of you—reunite in full resonance. When your frequencies merge, the star remembers itself."

Kael exhaled, eyes flickering with awe and disbelief.

"You're saying we're... supposed to recreate a star."

"Not supposed to," the Keeper murmured. "Destined to. And every time AstraCorps interferes, they reset the cycle. They kill one of you, capture the other, and the universe starts again."

Estelle's knees weakened. "That's why we've lived so many lives... why we've died before finishing what we started."

The Keeper turned away, shame crossing his face. "And my father helped make sure of it."

Silence fell again, broken only by the low hum outside, a vibration so deep it seemed to come from the soil itself.

Kael moved to steady Estelle, his hand warm against her back. "Then we finish it this time," he said quietly.

The Keeper gave a sad, knowing smile.

"Be careful what you finish. Stars don't burn without consequence. And once you begin, the universe will watch—and remember."

The lantern flickered once and went out, plunging them into darkness. The atmosphere outside swelled, low and alive, echoing through their bones.

Estelle's voice trembled in the dark. "Kael..."

"I know," he whispered back. "I feel it too."

The Keeper turned back toward them, the lantern's dying glow tracing the lines of age and regret across his face. He moved slowly to a drawer, pulling out a thin folder wrapped in weathered cloth. Inside were maps, notes, and photographs of constellations drawn by hand.

"I've spent my life tracking others like you," he said. "The fragments scattered across this world didn't all fall silent. Some found hosts—human bodies strong enough to hold their light. They don't remember who they are, not fully. They dream of stars they've never seen, hum melodies they've never learned."

He looked up at them with a faint, trembling smile. "You're not alone, Estelle. Not you, Kael. There are others—healers, empaths, visionaries. Each carrying a role in Lyratheon's rebirth."

Kael stepped closer. "And you've found them?"

"Some." The Keeper opened the folder, pulling out a single

sheet of faded parchment. "But there's one you must find first. He goes by the name Arion Vale. He's old now, older than I'll ever be, but he was the first Starborn to escape AstraCorps with fragments still alive in him. If anyone can help you understand how to survive what's coming—it's him."

He handed the file to Kael.

"He lives near the southern coast, on an island AstraCorps doesn't monitor. Coordinates are inside."

Estelle reached for the file, her hand brushing Kael's. The touch sent a familiar vibration through the room—a faint pulse that made the Keeper's eyes widen.

"It's starting again," he whispered. "The resonance. You're remembering each other."

Estelle's pulse quickened. "Remembering?"

"When fragments find their match, the memories return. That's what AstraCorps has always feared—because when you remember, you awaken. And when you awaken..."

The words trailed off as his expression shifted. His head tilted toward the door. "Someone's here."

Kael's instincts flared. He moved quickly, scanning the window.

A faint red dot pulsed across the curtain. Then another.

"Drones," he muttered. "They found us."

The Keeper grabbed Estelle's arm, his voice urgent but calm.

"Out back. There's a dock. Take the boat and follow the river south—it'll feed into the coast road. If you stay on land, they'll trap you."

"What about you?" Estelle said, panic rising.

He shook his head, smiling faintly. "I've lived my life running from ghosts. Maybe it's time I face them."

Kael gritted his teeth. "You'll die if you stay."

The Keeper's eyes softened. "Then I die giving your world a chance to live." He reached into his pocket, pressing a silver pendant into Estelle's palm. It shimmered faintly with a carved spiral—the same mark as the Lumen files.

"Keep this close," he said. "It's not metal. It's a piece of Lyratheon. It will remember for you when your memories begin to fade."

Estelle's throat tightened. "Keeper, we—"

He cut her off gently. "No. Just go. I've done enough harm in this life. Maybe it's time the next generation forgave what my father destroyed."

He turned toward the door as the sound of engines grew louder outside.

"Go," he said again. "Before they make you forget who you are."

Kael grabbed Estelle's hand and pulled her toward the back. They sprinted through the narrow hall, papers flying in their wake, the hum of drones overhead swelling to a deafening roar. They burst through the back door and onto the dock, the fog rolling thick over the dark water. A small wooden boat waited, its motor humming softly as if expecting them.

Kael helped Estelle in, untying the rope with one hand while keeping the other on her shoulder.

"Hold on," he said.

The motor kicked to life. The boat shot forward, cutting through the mist just as flashes of light split the trees behind them.

Estelle turned once, just in time to see the Keeper silhouetted in the cabin doorway, his figure framed by the approaching glare of searchlights.

He raised a hand in farewell.

Then the forest erupted in white.

CHAPTER 7
LILAC AND STARLIGHT

On the River

For a long while, neither of them spoke. The thrum of the motor blended with the rhythmic slap of water against the hull. The sky was an ink-dark canvas overhead, only the faint shimmer of starlight peeking through the fog.

Estelle clutched the pendant to her chest, her mind racing.

"He said it's starting again," she murmured. "The resonance. I thought it was just... connection. Attraction."

Kael's voice was low.

"It's more than that. It's memory."

She turned to him, her eyes wide.

"Do you think that's how AstraCorps finds us? Not by tracking us, but by tracking what we remember?"

He frowned. "What do you mean?"

"The frequency," she said, realization dawning in her tone.

"Our memories aren't just thoughts—they're energy signatures. When we start to remember who we are, we release radiation patterns. The same kind of readings AstraCorps calls anomalies."

Kael's grip on the wheel tightened.

"So the more we remember, the easier it is for them to find us."

She nodded slowly.

"And that's why the Keeper said it takes two. One Starborn alone doesn't have enough resonance to activate the frequency. But when we're together..."

Kael finished softly, "We become visible."

The river curved, the current growing stronger. Mist parted to reveal open water stretching into darkness.

"So they've been following me," Estelle whispered. "Tracking every dream, every flash, every time I feel something... familiar. They think I'm the anomaly."

Kael looked at her, a faint, rueful smile touching his lips. "Then they don't know what I am."

She stared at him, the invisible force between them

deepening. "Kael—"

He shook his head.

"Not yet. We'll figure it out when we find this Arion Vale."

Estelle looked down at the glowing spiral in her hand, her reflection flickering in the water.

"He said there are others like us. What if AstraCorps has them too?"

"Then we find them."

The wind caught her hair, sweeping it across her face as the boat carried them into the open current. Somewhere behind them, the horizon glowed faintly red—the dying light of the Keeper's cabin burning against the trees.

Estelle whispered, half prayer, half promise,

"We'll finish what he started. Lyratheon will live again."

Kael's eyes stayed on the horizon. "And this time, no one will destroy it."

Days Later

The morning news played on every screen across the

harbor café, the volume just loud enough to drown out the quiet conversations inside.

"Breaking story this morning: AstraCorps confirms a breach in its classified research division. Two fugitives have been identified in connection with the theft of restricted materials and biological data related to an important project. Authorities warn that the pair may be armed and dangerous. The individuals are believed to have ties to an eco-terrorist faction..."

Estelle stared at the screen, the words blurring as her face appeared—an ID photo from her lab badge, bright-eyed and harmless. Next to it, Kael's.

"'Threat to humanity,'" Kael muttered under his breath. "That's one hell of a promotion."

Estelle's stomach churned.

"They're framing us for their own crimes."

The café waitress passed by without looking at them twice, but Estelle still kept her head down. The pendant beneath her shirt pulsed faintly against her chest, as if echoing her heartbeat.

They'll track us through the cities," she whispered. "Every camera, every scanner."

Kael's eyes flicked toward the horizon.

"Then we disappear somewhere they can't reach. Somewhere that used to feel...safe" he said.

The drive was silent for miles, the hum of the engine the only sound between them. When the trees finally parted, Lake Meridian appeared like a memory – still, endless, the water mirroring the violet-gold sky. Kael slowed to a stop at the edge of the gravel drive. The cabin crouched beneath a canopy of pines, unchanged by time. He didn't move. His hands stayed tight on the wheel.

Estelle glanced over, her voice barely above a whisper.

"Are you okay?"

"I haven't been here since she died."

He didn't have to say who. Estelle already knew.

They stepped out. The air was cold, alive with the scent of pine, smoke – and lilac.

When Kael opened the door, time folded in on itself.

Everything was as it had been: the worn curtains, photo

frames softened by dust, the faint trace of cooking that clung to the air like memory.

Estelle followed quietly, letting him have his silence.

Kael brushed his fingers along the photographs on the mantle – his mother tending her garden, his smaller self holding a fishing pole too big for his hands. His heart began to race before his mind could catch up.

"She used to say the stars looked closer from here," he murmured.

"Maybe they were," Estelle said softly.

He nodded, eyes lingering on a single photo – his mother's gentle smile, her hand resting protectively on a boy's shoulder.

"Every summer she'd bring me here. Said it was safe. I didn't know what she meant until now."

Down the narrow hall, the smell of her perfume seemed to linger. He pushed open his old bedroom door; the hinges creaked like a familiar voice.

On the bed sat a single box, untouched and waiting.

"She must've left it," he whispered.

He brushed off the dust and lifted the lid.

Inside were photos – flickering like captured starlight: him laughing beneath a canopy of fireflies, her sitting on the porch, face lifted toward the stars, the two of them in a garden, sunlight tangled in her hair.

At the bottom, beneath a folded lilac-scented scarf, lay an envelope.

For Kael – when you're ready.

Her handwriting flowed steady and sure. He unfolded the letter.

My dear boy,

If you're reading this, I'm gone. That's the only way I'd ever let you find this box.

There are things I never told you – things I couldn't. I worked once for people who studied the light in ways they shouldn't have. I thought we were learning how to harness energy from the stars. I was wrong.

What we studied wasn't energy; it was life. Children. You were one of them, Kael.

They called you a "subject," but to me, you were a

miracle. You survived what none of the others could.

erased your files. I ran. For years, I prayed they'd never find you. You were never born of me, but I have never loved anything more fiercely. You were my heart. And that is enough for any mother.

They may come for you someday, but I pray you'll be strong enough to find where you came from. And I pray you'll find her — the other light.

You'll know her by the way the world hums when she breathes.

Always,

Mom

Kael's hands trembled. He read it once, then again. The word children echoed in his skull.

"She knew," he whispered. "She knew everything."

Estelle's hand found his arm.

"She didn't just know," she said. "She saved you."

He nodded slowly. "All this time, I thought she just... saw something in me."

"She did," Estelle said, smiling faintly. "She saw the stars."

He let out a shaky laugh. "I can't tell if that makes her crazy or divine."

"She made you human," Estelle said gently. "That's divine enough."

Silence fell between them, soft and warm. Outside, the first stars blinked over the lake—reflected in the ripples like scattered light.

For the first time since AstraCorps, they weren't running. They just were.

Kael tucked the letter into his jacket and stepped out toward the dock. Estelle followed.

"She worked for them," he said at last. "AstraCorps. I found her badge. She helped build the same system that hunted us."

Estelle looked out over the water. "You blame her?"

"I don't know. I wish I could. It'd be easier than wondering if she ever regretted it."

"She did," Estelle said. "You said she loved you?"

"With everything she had."

"Then that's all that matters."

"I was adopted," she added softly. "My parents weren't bad, just... distant. They tried to love me through silence. But maybe your mother—seeing what AstraCorps was doing—maybe it helped her understand you more. Helped her love you harder. Sometimes, knowing darkness makes us fight for the light."

Kael's breath hitched into a laugh. "You make it sound poetic."

"It is," she said. "You are."

Their eyes met, and the hum that had haunted her dreams rose between them – soft, familiar, alive.

"Estelle," he breathed, her name a prayer.

Then she was in his arms, and the night opened.

Their lips met – slow, inevitable, electric.

Above them, the sky responded.

Clouds parted, stars brightened, and the constellations shimmered in rhythm with their heartbeats – two lights finding their orbit again. When they pulled apart, the air glowed faintly around them.

"What was that?" Kael whispered.

Estelle smiled, tears shining in her eyes. "Home," she said. "That's what it feels like."

Later, they built a fire near the shore.

Estelle sat wrapped in one of his mother's handmade blankets, a mug of hot chocolate warming her hands. The reflection of the flames danced across the lake.

"I used to think I joined the Air Force to see the world," Kael said. "But after her death, I think I stayed in because I wanted to feel closer to her. To the sky."

Estelle leaned against him. "She never really left," she said. "You just stopped looking up."

He chuckled softly. "You sound like her."

"Maybe she's speaking through me."

He turned toward her, the firelight tracing her face in gold. She reached for his hand; their fingers interlocked.

The air pulsed. A shimmer rippled over the water.

He caught his breath. "Did you feel that?"

"Yes."

"Like the world just... breathed?"

She smiled. "Maybe it did."

The lake shimmered brighter, twin reflections of light intertwined on its surface.

The Echo of Lilac

When Kael finally slept, Estelle wandered to the water's edge. The air smelled faintly of lilac again – sweet, impossible, timeless.

The hum began, wrapping itself around her heartbeat. She looked down and saw a woman's reflection in the rippling water – kind, luminous, smiling.

"You were the safe haven," Estelle whispered. "For all of us."

The reflection shimmered, as if in acknowledgment.

Behind her, Kael stirred.

"Were you talking to someone?"

She smiled softly. "Just to your mother."

He nodded, half-asleep, as though that made perfect sense.

Above them, the constellations shifted – forming, for the briefest moment, a glowing figure-eight in the sky.

A promise.

CHAPTER 8
OTHERS LIKE US

They drove for hours along the southern coastline. The landscape shifted from cracked highways to open wetlands, where reeds swayed in the wind and the ocean spread wide like melted silver.

Estelle stared out the window, the trees blurring into streaks of green and gold. Kael glanced over, catching the worry in her eyes. He reached across the console, his fingers finding hers.

"Hey," he said gently. "Everything's going to be okay."

She didn't answer, but her grip tightened.

For a long moment, the hum of the road filled the silence – steady, grounding. Both of them had grown up as only children, their worlds built from solitude. Kael had his blueprints and broken engines; Estelle had her sketches of stars that no telescope had ever seen.

But now, as their hands remained intertwined, that quiet loneliness that had once defined them felt distant – almost forgotten.

For the first time, they weren't alone.

Kael smiled faintly as the road curved ahead. "Almost there," he murmured, nodding toward the faint light breaking through the trees.

Estelle followed his gaze.

By dusk, the coordinates the Keeper had given them led to a secluded inlet surrounded by moss-draped oaks. A single structure sat on the water's edge. Smoke curled faintly from the chimney.

"I just hope we find the answers we've been looking for," Estelle whispered, her voice barely rising above the sound of the engine.

Kael glanced at her – really looked at her – and for a heartbeat, it felt as if she were searching his soul for a truth he wasn't sure he could give.

"We will," he said, his tone steady but threaded with hope.

The air felt different here–thicker. Estelle could sense it the moment her boots hit the ground. The vibrations of the frequency that had haunted her for months dulled to a low, steady throb.

She glanced at Kael. "You feel that?"

He nodded. "It's quiet. Like the noise finally stopped."

"Then this is why they've never found him," she said softly. "The frequency's muted here. AstraCorps can't trace it."

They approached the cabin carefully. The porch creaked under Kael's weight as he knocked.

No answer.

"Maybe he's not home," Estelle whispered.

But then the door creaked open on its own.

Inside, the room smelled of salt and cedar. Dust danced in the thin beam of light cutting through the shutters. And there, in the center of it all, an old man sat in a rocking chair, eyes closed, hands folded in his lap. He looked nearly lifeless—his skin paper-thin, his frame brittle.

"Arion Vale?" Kael asked.

The man's eyes fluttered open. For a heartbeat, they were clouded with age. Then he saw them.

The rocking stopped. His pupils flared gold.

He rose—slowly at first, then suddenly with strength that shouldn't have been possible—and dropped to his knees.

"It's you," he breathed. "The twin stars. The heartbeat of Lyratheon."

Estelle froze. "Please, don't—"

But Arion bowed deeply, trembling. "We've waited lifetimes for your return."

Kael helped him up gently. "We don't understand any of this. The Keeper sent us—he said you could help."

Arion smiled faintly, lines of age softening around his mouth. "He always was a believer in redemption. I suppose we both owe you the truth."

He gestured for them to sit. As they did, a faint light began to shimmer beneath his skin—subtle, like fire beneath frost. His voice grew stronger with every word.

"Before AstraCorp ever touched Lyratheon... our world was already dying."

Estelle frowned, but he continued, gaze distant.

"It wasn't war. It wasn't sickness. It was the core—our

heart. It was fading. So our two young leaders, barely twelve, chose the unthinkable. They left Lyratheon to merge their cores… to reignite the world from within. A sacrifice no child should have made."

He paused, and something like grief flickered across his face. "As a failsafe," he said quietly, "their celestial essence was anchored to fragments of Lyratheon. If the merge failed, their souls would return to those fragments and be reborn… again and again… until the mission succeeded."

Estelle felt the weight of it settle in her chest.

"For a few years, we waited," Arion whispered. "Believing they would return. Believing Lyratheon would rise again."

His jaw tightened.

"But AstraCorp found us first. They stripped our world for its light… and then destroyed what remained." His voice wavered, just once. "After that, I nearly gave up hope entirely. With Lyratheon gone, the children had no home to return to—not until their mission was complete. Their rebirth scattered across the stars."

He drew in a slow breath and met her eyes—steady,

reverent, certain.

"I have searched a lifetime for the two who would finish what they began. For the ones destined to restore what was taken from us." His voice dropped to a quiet hush. "...and now you stand in front of me. The Starborn weren't meant to be soldiers. We were meant to be guardians—keepers of balance between worlds. When Lyratheon fell, many of us fell with it. Those who survived carried fragments of the star inside them. Some Astraborn, like the Keeper, tried to hide us. Others... surrendered to fear."

He glanced toward the doorway. "Jade, you can come out now."

From the shadows emerged a woman—tall, fierce, eyes like burnished bronze. She wore a tactical harness over weathered leather, a blade to the left of her hip and a sidearm at her right. Her presence radiated control.

"Been listening for the last ten minutes," she said flatly. "They talk too loud."

Kael bristled, but Estelle caught the faint curve of a grin on the woman's lips.

"This is Jade," Arion said. "She was one of the first I found. Or rather, she found me—dragged me out of a fire I set myself."

Jade leaned against the wall, arms crossed. "He's being modest. He was half-dead and ready to let AstraCorps take him. I just gave him a reason to stop whining."

Arion chuckled softly. "She's been keeping me alive ever since."

Estelle studied her. "You're Starborn too?"

Jade shrugged. "Half. Fragmented, like most of us. My light's a little dimmer than yours, sweetheart, but I make up for it in personality."

Kael smirked despite himself. "So there are others?"

Jade's expression hardened. "Plenty. Some of us fight back. Others... forgot what they are. AstraCorps has them now. Brainwashed. Fed lies about what Lyratheon was. They think we're the threat."

Arion's face grew somber. "Not all Starborn remember their purpose. Some believe AstraCorps's vision of 'containment'—that merging worlds would destroy humanity. Those are the ones you must fear most."

Estelle frowned. "And you? You've been protecting the rest?"

He nodded. "Jade and I built a sanctuary here. The resistance—those who remember—gather nearby. This place dampens our frequencies. AstraCorps can't trace us here. We use that to find the others before they do. I have something I'd like for you to see," Arion said, his smile as wide as the night sky.

Jade moved to his side, steadying him as she pressed her palm against a star-shaped lantern fixed to the wall. The metal warmed beneath her touch, and with a low hum, the wall shifted open, revealing a hidden passage.

They stepped inside.

The elevator descended in silence, the hum of energy vibrating faintly through the floor. Estelle watched the light flicker against the steel walls, each pulse syncing with her heartbeat. When the doors slid open again, cool air swept across her face, smelling faintly of stone and ozone. They followed a narrow corridor lined with cables and softly glowing symbols until they reached a massive steel door. A man stood guard – tall, pale, his long coat brushing the floor.

The moment they approached, he inclined his head, voice low and reverent.

"The air hums different when you're near," he said. "Welcome home, both of you."

The door hissed open.

Estelle glanced at Kael, unease rippling through her chest. "Home?"

Arion only smiled, stepping aside. "You'll see."

They stepped through.

The space opened into a vast underground dome – the heart of the resistance.

A glass arch crowned the ceiling, revealing a sky full of stars that seemed impossibly close. The air vibrated with quiet energy.

Men and women moved about the hall – some hunched over glowing maps, others sorting equipment. Laughter echoed softly as children darted between them, their skin faintly illuminated, leaving trails of starlight that shimmered and vanished like comet tails.

Estelle froze. Their energy – their presence – felt like her own. When the murmurs spread that they had arrived, the hall went still. One by one, the Starborn rose to their feet.

"It's them," a voice whispered. "The Twin Cores."

Arion raised his hand, and the sound swelled – not applause, but something deeper. The hum. Dozens of frequencies converging until the air vibrated like a single living heartbeat.

Kael's instinct was to retreat, to tell them he wasn't anyone's savior. But Estelle's hand brushed his. He stayed still.

Arion stepped forward, his voice carrying the calm authority of someone who had waited a lifetime for this moment.

"For years, we've hidden, scattered across the world – remnants of a light that refused to die. But the prophecy said the two hearts of Lyratheon would one day awaken, reborn among us. That when they found each other, their

union would call the stars home again."

A hush fell.

"And when their hearts merge," Arion continued, "a third shall rise – a bridge between worlds. The child who will mend what was broken."

Estelle's pulse quickened. She looked at Kael – but his eyes were fixed on Arion, unreadable.

"I don't want worship," Kael said. "I just want answers."

Arion's smile was kind. "You'll have them. But first, you must see who you're fighting for."

He gestured toward the far side of the hall, where families had gathered. Children with faintly glowing freckles clung to their parents' legs. A young boy traced constellations on the floor with his fingertip – the same figure-eight Estelle had drawn since childhood.

Her throat tightened. "They're... like us."

Arion nodded. "Others like you. Some born of the experiments. Some born free. All waiting for the day the stars return."

The hum rose again, softer this time – a sound that felt

like gratitude.

Estelle swallowed hard. "You've kept them safe?"

"As best we can," Arion said. "But AstraCorps hunts us relentlessly. They fear what they don't understand – what they can't control."

Kael's jaw tightened. "And you think we can stop them?"

"I think you're the reason we still believe we can," Arion replied.

Later, Estelle found herself sitting by one of the smaller fires outside the main hall. The night was cool, filled with murmured voices and the distant laughter of Starborn children.

A woman settled beside her – older, her dark hair streaked with silver light that shimmered in the flames. She introduced herself as Nerai, one of the first Starborn to find Arion.

"They look at you like you're holy," Nerai said gently. "Does it frighten you?"

Estelle stared into the fire. "It terrifies me. I don't feel holy. I barely feel like I belong here."

Nerai smiled. "That's how we know you do. The true ones never arrive certain of their light."

Estelle looked at her, brow furrowed. "You believe the prophecy – the one about the two hearts and the child?"

"I believe it because I've seen the signs," Nerai said. "The stars dimmed when you were apart, and they began to flicker again the night you found each other. That kind of balance isn't coincidence – it's cosmic memory."

She reached out, touching Estelle's hand. "When the hearts merge, the energy will renew Lyratheon. But the prophecy doesn't just speak of rebirth – it speaks of choice. You and Kael will have to decide if you'll remain here, or if you'll return home."

Estelle's gaze drifted toward the children playing in the field – laughter like wind chimes, starlight in their eyes. "How could we leave them?"

Nerai's voice softened. "You won't leave them. You'll lead them."

Estelle's chest ached. "But what if we fail?"

Nerai smiled faintly. "Then we keep the faith until the

stars rise again."

As night deepened, Kael joined her by the fire. He glanced toward the sky – now alive with a thousand faintly glowing lights that hadn't been there before.

"Do you see that?" he whispered.

She nodded. "They're coming back."

Kael exhaled slowly. "Maybe it's us," he said. "Maybe we're pulling them home."

And for the first time, Estelle didn't deny it.

All around them, the Starborn sang – a low, resonant hum that seemed to thread through every living thing.

Estelle leaned against Kael's shoulder, her eyes fixed on the constellations overhead.

The stars no longer looked distant.

They looked like they were listening.

The next morning, soft light spilled through the cabin's wide windows, painting the floor in shades of gold and blue. The world outside was quiet – a rare peace after the endless running.

They met Arion inside his cabin, the air still scented faintly *of cedar and smoke. Estelle stood near the window, watching a group of Starborn children chase one another. For the first time, she didn't feel like an anomaly.*

She felt like part of something – ancient, inevitable.

Purpose no longer felt like a curse.

Kael felt it too. The night before had stripped away any doubt that what they'd found here – what they were – was real. The hum beneath his skin was steady now, not wild or untamed, but aligned. He remembered his mother's letter, her words about leading when the time came.

And now, standing here with Estelle beside him, he finally understood.

Arion sat at his desk again, eyes scanning a digital map that pulsed with faint lines of light. Near the door, two worn travel packs waited, their straps neatly buckled.

Kael nodded toward them. "I'm assuming Jade has a mission?"

Arion's lips curved faintly. "Indeed. As I've said before, we must bring together the remaining Starborn before AstraCorps does. Many are still in hiding – frightened, scattered. But they'll come when they know the Twin Cores have returned."

Kael leaned forward, forearms resting on his knees. "The Keeper said you were tracking someone before we arrived."

Arion's gaze flicked to him, then to Estelle. "Yes," he said. "Another Starborn. Young. Unstable. Their energy's flaring faster than anything I've ever seen. I was just sending Jade to find them."

Estelle tore her gaze from the window, meeting his eyes. "Then tell us where to start."

Jade's smirk returned. "Guess I'm getting company then."

Arion placed a hand on Kael's shoulder. "I cannot go with you. My time here is almost done. The light fades with distance, and I must protect what's left of this refuge. But Jade can guide you. She knows the paths AstraCorps doesn't."

He looked between them, his expression both proud and mournful. "You two... you're the reason we've survived this long. The fragments inside you are the core of Lyratheon itself. When you merge again—truly merge—the star will live. But until that day, you must gather our brothers and sisters. Bring them home."

Estelle nodded slowly, her throat tight. "We'll find them. All of them."

She was no longer running from what she was.

And neither was Kael.

Jade slung a bag over her shoulder, glancing between the two of them. "Then we move now. AstraCorps won't stop hunting you—and if they find this place, everyone here dies."

Kael met her gaze. "We'll keep them safe. Whatever it takes."

Arion smiled faintly, already looking younger—the glow beneath his skin soft and alive.

"The universe finally remembers," he whispered. "Now it's time humanity learns to do the same."

Later that night, Estelle stood on the shoreline outside the cabin. The sea shimmered faintly, starlight rippling across its surface. Kael joined her, his presence quiet but grounding.

"You think we can really bring them all together?" she asked.

Kael looked at the stars. "We already are."

Behind them, inside the cabin, Arion watched from the window, his reflection glowing brighter with every breath he took. Jade cleaned her weapons, muttering under her breath. For the first time in years, the air around them thrummed not with fear—but with purpose.

AstraCorps Headquarters – Interrogation Room 3

The walls were white, too white, humming with the faint vibration of cooling vents and hidden cameras.

Sori sat in a metal chair under a single light, her palms damp, her heartbeat too loud in her ears.

Across from her stood Director Voss—tall, gray suit, colder eyes. Nora Wimbush paced behind him like a vulture in heels.

"Dr. Cho," Voss said evenly, "you were close to Estelle Auryn. Has your assignment been compromised?"

"No," Sori replied, her voice steady despite the pressure in *her chest. "We worked together. She was brilliant. Impulsive, but brilliant."*

Voss slid a photo across the table—Estelle's face on a security feed, her hair whipped by wind, Kael beside her. "You're aware she and Captain Verran have stolen highly classified materials."

Sori hesitated just long enough to be believable. "I heard the rumors. But I don't believe she'd betray AstraCorps."

Nora stopped pacing, her mouth curling into a sharp smile. "Believe it, sweetheart. She and your pilot boy cost us three satellites and a whole team of analysts. If they're caught, it's not just them on the line—it's everyone who touched that project."

Voss leaned forward, his voice almost gentle. "You're one of the few who knew her personally. That makes you valuable. Tell us where she'd go."

Sori lifted her gaze to meet his. "If I knew, I'd already be there dragging her home myself."

A long pause. The sound of the air vents filled the silence.

Finally, Voss straightened. "Keep your comms open. You're

on internal review until further notice."

Nora leaned down, her perfume sharp as acid. "You think she was your friend," she whispered. "But if you want to keep breathing, you'll help us find her."

The door opened. Guards escorted Sori out. Her hands shook only when she was alone.

In the elevator, she pressed her forehead to the wall and whispered, "Hang on, star girl. I'll stall them as long as I can."

Coastal Highway – Two Weeks Later

The open road stretched ahead like a ribbon of dust and heat. Estelle sat in the back of the jeep, watching the world blur by—fields, forests, rivers that glimmered like veins of light.

Jade drove with one hand on the wheel, sunglasses hiding her eyes. Kael rode shotgun, scanning the horizon for drones.

They'd been moving for weeks—sleeping in abandoned motels, barns, even caves. Each night, the hum inside

Estelle grew stronger. Her body buzzed with memory; her dreams burned with flashes of galaxies and laughter not of this Earth.

Jade called it "resonance sickness."

Kael called it destiny. And one night, somewhere between exhaustion and eternity, it pulled them together.

That night they stopped at an old wind farm on the edge of the desert, where turbines turned lazily against the indigo sky. The stars above were endless—too bright, too close. The hum between them both until it felt unbearable.

Kael reached for her hand, and she didn't pull away.

"You feel that too," he said softly.

She nodded. "Like the whole universe is waiting for something."

Then he kissed her. It wasn't gentle. It was inevitable. A collision written into the fabric of things long before they were born. The air shimmered around them, particles vibrating in gold and blue. Their bodies moved in rhythm with something larger than themselves—every breath a pulse of light, every heartbeat a spark. The world tilted, and for one perfect moment, they were weightless.

And then, a piercing whine split the silence.

Estelle gasped, rolling away. The frequency reader she carried in her pack blinked violently, its screen flooded with impossible numbers.

"They'll find us," Jade said from the jeep, her voice sharp, disgust edging her tone. "If you two decide to make love again, do us all a favor and do it underground."

Kael pulled on his shirt, half laughing, half terrified. "How bad?"

"Off the charts," Jade snapped. "You basically lit up every satellite within two hundred miles."

Estelle's cheeks flushed, but her expression was grim. "Then we move now."

By dawn, they reached the foothills of a ruined mining town—the coordinates Arion had given them for the next Starborn. The air tasted of dust and iron.

Jade scouted ahead while Estelle and Kael checked the equipment. The hum of the frequency reader had calmed, but not completely.

"He's close," Estelle whispered.

Kael nodded, hand hovering near his weapon. "Or waiting."

They entered an old refinery at the edge of the town. Broken skylights spilled thin shafts of morning light across rusted beams and puddles of oil.

At first, it was silent. Too silent.

Then the sound of boots.

Dozens of them.

"Run," Jade hissed.

The world erupted in light. Drones descended through the roof, beams scanning. AstraCorps insignias gleamed on the armored suits pouring in from every entrance.

"Go!" Kael shouted, shoving Estelle toward the back.

Jade dove into the shadows, her silhouette vanishing as she pulled a flash charge from her belt.

Kael fired at the drones, hitting one, then another, sparks raining from the ceiling.

But Estelle froze. They weren't aiming at her. They were looking at her.

She saw her own reflection in their helmets—her face, illuminated by the faint gold glow under her skin.

"They already know who I am," she said softly.

Kael turned, panic in his eyes. "Then we fight!"

She shook her head. "No. They don't know you. That means you're still safe."

"What the hell are you talking about?"

She reached into her pocket, pulling out a small vial filled with shimmering liquid. Kael's breath caught.

"That's the metal from the pendant," he realized. "You melted it."

She nodded, tears in her eyes. "It holds our frequency. Our memories. If they take me, I'll lose everything. But if you drink it... it'll remember for both of us."

He shook his head. "No. We do this together."

She pressed a trembling hand to his cheek. "You'll find me. Save me when we have what we need."

"Estelle—"

Before he could stop her, she swallowed the vial's contents. The glow beneath her skin flared, burning bright, then dimming to nothing.

"Find me," she whispered. "Always."

The soldiers swarmed in. Kael lunged forward, but the blast of light knocked him back, searing through the air like a flare.

When he opened his eyes, Estelle was gone.

Only the faint hum of her frequency remained—

a ghost note fading into silence.

Somewhere deep underground, AstraCorps locked onto a new signal.

And far above, in a forgotten lab, a monitor flickered—

Subject Recovered. Memory Containment: In Progress.

CHAPTER 9
THE TRACKER

Sori

When Sori was a little girl, she thought the stars were the souls of people who'd done something right.

Her mother told her stories like that – soft, simple ones meant to help her sleep. Her father told her something different.

"The stars aren't wishes, Sori-Ana. They're orders waiting to be followed."

She used to sit beside him on their apartment balcony in Seoul, her small fingers clutching the edge of his old telescope as he adjusted the lens. His uniform gleamed even under the weak city light, medals heavy on his chest. When she looked up, she saw her reflection in his polished brass buttons – wide eyes, full of galaxies.

He worked for the Republic's Space and Defense

Commission, one of the youngest military liaisons ever appointed to the international task force that would eventually become AstraCorps. Sori idolized him. The way he stood. The way people listened when he spoke. The way the sky seemed to bend for him when he pointed to it.

So she decided she would grow up to follow him – not just in rank, but in purpose.

She'd build machines that reached the stars.

She'd make him proud.

By ten, she'd memorized every constellation visible from their rooftop.

By thirteen, she was winning regional science awards.

By twenty, she graduated top of her class – aerospace systems and astrophysics, Seoul National University.

The top was everything she'd imagined.

And exactly as lonely as they said.

People admired her work ethic, her focus, her flawless manners. But admiration was not friendship, and excellence was not love.

Sometimes, after long hours in the lab, she'd catch her

reflection in the glass – tired eyes, dark circles – and imagine her mother's gentle laughter.

Her mother had been warmth, chaos, heart.

A jazz singer from Atlanta who met her father on a diplomatic trip.Together they were lightning and lullabies – until lightning won. Her mother died when Sori was twelve. After that, silence became law in their household. No jazz, no emotion, no softness.

"Discipline, Sori-Ana," her father would say, pressing another textbook into her hands. "The world doesn't need your feelings. It needs your results."

She learned to bury her feelings deep – so deep she sometimes forgot they existed.

When AstraCorps International offered her a position years later – an accelerated track in defense technology and energy research – she didn't even hesitate.

Her father's handshake that day had been the proudest moment of her life.

And the loneliest.

Because for all her brilliance, all her ambition, she didn't *yet understand what kind of stars she was helping to reach for.*

The scent of barley tea drifted through the cold metal corridors of AstraCorps' central tower – soft, warm, and deeply out of place.

Sori-Ana Cho cupped the mug in both hands, the way her halmeoni had taught her, as if warmth could erase guilt. Steam brushed against her cheeks, blurring the reflection in the window – a tired woman in a lab coat too white, eyes ringed in sleepless gray.

The city below was all glass and order. It was the kind of structure her father had spent his entire life worshiping. "Discipline brings honor," he used to say in his crisp Seoul accent. "Emotion builds chaos."

She'd joined AstraCorps because of him.

Because the Cho family name had meant something – once.

Because her father had been a defense attaché before defecting to the U.S. with nothing but his medals and regret. He'd wanted his daughter to rebuild their legacy through obedience. Through science. Through service.

She never meant for that obedience to cost her soul.

Now, as she stared at the live feed flickering on her monitor — Estelle Auryn, asleep under dim lab light, her breathing steady, her face peaceful — Sori's stomach twisted.

Her friend.

Her sister.

Her undoing.

They had been so alike, two only children orbiting loneliness.

Sori remembered their first late-night study session in grad school — sharing instant ramen and laughter over failed experiments.

She'd been the first to call Estelle Star Girl, teasing her for daydreaming in data sheets.

Now, that same girl was listed as Subject 9.

She ran her thumb along the mug's rim, whispering a small Korean prayer her mother used to say before bed.

Bokjareul gidohae.

Pray for forgiveness.

The door slid open with a hiss.

Director Voss filled the doorway, immaculate as ever – not a hair out of place, not a speck on his black uniform. His presence made even the lights seem sharper, crueler.

"Sori-Ana," he said, his tone smooth and cordial. "You've been here since dawn."

"Couldn't sleep," she replied, setting the mug down. "The frequency spikes haven't stabilized since—"

He cut her off with a raised hand. "We'll handle the anomalies later. Sit."

She obeyed. Voss walked to the window, hands clasped behind his back. His reflection looked like a statue carved from shadow.

"You've spent more time with Subject 9 than anyone," he said. "You've studied her responses, her fluctuations, her emotional patterns."

Sori swallowed hard. "She's not a subject. Her name is Estelle."

Voss smiled faintly. "That's the problem with you, Dr. Cho.

You get sentimental."

"Maybe that's what makes me human."

"Human," he repeated, rolling the word like it offended him. "Do you know what humans do when they're afraid? They destroy. That's what the Starborn will do once they realize what they are. They'll rise. They'll burn this world to cleanse it."

"That's not true." Her voice trembled. "You don't know them."

"I know what they are," Voss said sharply. "Two cores – merged by design, not chance. They were created to consume energy, not to preserve it. You've seen the readings. Every time they connect, the power grid falters. The atmosphere bends."

He leaned closer, voice low and cold.

"Do you think the Earth can survive such enegy walking its surface again?"

Sori gripped the edge of the console. "They're not threats."

"But they are," he said softly. "The worse kind. At least if they were human, they could understand restraint. You;re

a woman of science Dr. Cho, what does the numbers show?"

Her throat closed. She wanted to scream, to throw her mug, to tell him he was wrong – but the logic in his tone was disarming. It sounded like protection, not control.

When he left, the hum of the servers filled the silence. Her reflection in the glass looked smaller, lonelier.

The next morning, her inbox pinged – another message from Nora Wimbush.

From: N. Wimbush

Subject: TRKRR Report Deadlines

Sori-Ana, *you're a bright girl. I do not need to remind you of company deadlines for* TRKRR *reports. Do not let your emotions get the best of you.*

Her hands trembled as she closed the email. Nora's *words burned more than they should have.*

Sori *stared at* Estelle's *photo on her monitor – laughing mid-sentence, eyes bright, curls framing her face. She had light. Not the one* AstraCorps *wanted to dissect – the one that made you believe people were still good.*

The closer Sori got to her, the harder it became to report.

Every observation felt like betrayal.

Every check-in felt like a confession.

And yet, she couldn't stop.

AstraCorps didn't allow absences in loyalty.

That night, Sori watched Estelle's biometric data flicker across her screens – the rhythmic heartbeat, the steady breathing, the faint glow under her skin.

She hesitated for a long moment before opening the encrypted channel marked TRKRR_04: 09 ACTIVE STATUS.

A single tear fell onto her keyboard.

She whispered, almost too softly to hear:

"Mianhae, eonni. Forgive me, sister."

And then, with shaking hands, she pressed send.

AstraCorps Three Days Later

The corridor lights dimmed as alarms whispered in the distance – not danger alarms, but the kind that signaled a new delivery. Sori's pulse quickened. She turned toward the observation bay, barley tea still cooling on her desk.

Two guards in matte-black suits wheeled in a stretcher.

On it, Estelle lay motionless.

Her hair fanned across the pillow like spilled ink, skin pale under the harsh fluorescence. Electrodes dotted her temples.

A small tremor went through Sori's hands.

She set down the mug before it shattered.

"What are you planning to do to her?" she asked, her voice barely steady.

Nora Wimbush didn't even look up from her tablet. Her lipstick was perfect, her tone colder than the steel walls around them.

"Whatever is necessary," she said. "Director Voss wants her prepped for cognitive isolation and resonance stripping by morning. The subject's energy levels are... inconveniently high."

"Inconveniently?" Sori echoed, disbelief cracking through her voice. "She's human."

Nora's gaze lifted, sharp and amused.

"She stopped being human the moment the readings hit 9.7. You know the protocol, Dr. Cho. You were briefed on Project Lumen."

The name hit Sori like a blade to the gut.

She did know.

She knew exactly what happened to defected Starborn – how they were restrained, drained, rewritten until obedience was the only thing left in their eyes.

She'd read the reports that never saw daylight.

She'd coded the locks that kept the subjects from screaming too loudly.

And now it was Estelle.

Her sister in every way that mattered.

A second stretcher rolled past, this one guarded more heavily.

The figure strapped to it was a man – tall, copper-skinned, restrained at the wrists and throat. His pulse readings flickered violently across the monitor.

Voss's voice echoed from behind the glass as he approached the chamber.

"Keep him heightened. The resistance is good for yield. If we can harness that much celestial energy without combustion, he'll power the first generation of our stellar weapons."

Sori's breath caught. Weapons.

She turned back to Estelle's still body, a thousand thoughts colliding in her chest.

Whatever AstraCorps was building, whatever salvation they claimed to serve – it had nothing to do with science.

And everything to do with control.

Nora scribbled something on her datapad, her voice clipped.

"You should go home, Sori-Ana. You look tired."

But Sori didn't move.

She just stood there, hand pressed against the glass between them, watching the faint glow pulse under Estelle's skin – the last flicker of light in a place that had forgotten how to feel.

For the first time since joining AstraCorps, Sori stopped taking orders.

Voss

Director Voss stood behind the glass, hands clasped behind his back, the faint hum of containment fields vibrating through the floor.

Estelle Auryn's vitals glowed steady on the monitor – an impossible reading for a human, radiant even in sedation. He studied the pulse like a priest reading scripture. He wasn't proud of what he'd done. Pride was for men who still doubted their purpose.

He'd seen the simulations – the heat maps, the projections, the long decay. Earth was dying. Not today. Not tomorrow. But soon. Its atmosphere bleeding thin, its core spinning slower each decade. Humanity would never admit it. They'd blame climate, politics, God. But Voss had seen the truth written in light.

The Starborn energy could reset the planet's failing magnetic field, if harnessed correctly. Controlled. Contained.

Unrestrained, it would tear reality apart.

"They think they're gods," he murmured to himself,

watching Estelle twitch faintly under the electrodes.

"But I'm the one trying to save the world from them."

The assistant at his side hesitated. "Sir, the resistance—"

"—is irrelevant," Voss interrupted. "We can't afford mercy. We are out of time."

He reached for the datapad and traced the readings – the hum rising, the luminous energy growing brighter than the fluorescent lights above.

For a fleeting second, he thought it was beautiful.

Then the monitor flashed red.

He smiled faintly. "Yes," he whispered. "Burn bright. For all of us.

He had a sinister grin accross his face. He looks at the other scientist "And Dr. Foreman, prep the new starborn for celestial extraction"

CHAPTER 10
FOR THE GOOD OF HUMANITY

Estelle

Light.

White, sterile, endless.

For a moment she thought she was still dreaming, drifting inside one of the fractured memories that came like storms —Kael's hand reaching for her, the smell of salt, the hum of stars inside her chest. But when she tried to move, the restraint bands bit into her wrists.

A soft mechanical hiss answered her breath.

Somewhere nearby, monitors pulsed in rhythm with her heartbeat.

Three months.

That's what the wall calendar said in red marker when her vision steadied. Three months since the trap.

Footsteps clicked against tile.

"Welcome back to the land of the measurable," said a voice too smooth to belong to comfort.

Nora Wimbush stepped from the shadows, lab coat immaculate, lips curved into something between a smile and a sneer.

Estelle tried to sit up. "Where am I?"

Nora pressed a button; the bed's restraints retracted halfway. "Home. Or the only one that will ever tolerate you. AstraCorps."

Memories stabbed through her like broken light: the refinery, Kael's face, the burst of energy. Then blackness.

"Kael..." she whispered.

Nora's smile sharpened. "Ah, yes. The pilot. Don't worry—our retrieval teams are still looking for him. Though honestly, we're more interested in you."

She approached the bedside, heels echoing against steel. "You're one of the twin cores which makes you special only for science. But you are one of many Starborn that we have in custody. We ran some tests while you were sleeping."

Estelle's pulse raced. "Tests?"

Nora turned a monitor toward her. A grayscale image glowed on the screen—two flickering heartbeats, one inside the other.

"Congratulations," Nora said lightly. "You're expecting."

For a second the world tilted. Estelle's breath caught. "No. That's impossible—"

"Oh, it's very possible." Nora's tone was almost gleeful. "Apparently cosmic biology is quite... fertile. You and your pilot managed to give us something priceless while you decided to play Bonnie and Clyde across state lines."

She leaned closer, her perfume sharp, clinical. "We plan to take samples of the fetal tissue. Early enough to replicate the DNA without the complications of consciousness. With luck, we can create our own Starborn line—without the inconvenient free will."

Estelle's vision blurred with fury. "You can't—"

"Oh, we can." Nora straightened, voice turning to ice. "And before you waste breath asking where your friend Dr. Cho is... she's the reason we found you."

Estelle froze. "Sori...?"

Nora nodded. "Tracker designation 4-B-3. The name 'Cho' has no significance. You honestly didn't notice her surveillance? They never do."

She began circling the bed, her words slicing like glass.

"Trackers are inserted into Starborn lives early—mentors, lovers, friends. Their purpose is simple: keep the subject calm, compliant, unaware. The moment you started digging, she reported it."

"No." Estelle shook her head. "Sori would never—"

"She did." Nora's smile softened into mock pity. "And you know what's remarkable? She actually cried when we brought you in. Almost like she forgot what side she was on."

Estelle turned her face away, tears burning hot. The hum in her chest—once a comfort—now a scream.

Nora tapped the monitor off. "You're not special, Estelle. The world just treats you that way because it's afraid of you. But fear can be repurposed. Quantified. Reproduced."

She gestured to the door. Two guards entered, silent as machines.

"Prep her for Phase Two," Nora ordered. "We start sampling in forty-eight hours."

As they wheeled Estelle down the corridor, the lights above flickered—once, twice—then steadied. The sensors on the wall registered a faint surge of electromagnetic energy.

Somewhere deep inside her, the star refused to stay quiet.

Kael

Rain hammered the windshield as the jeep tore through a stretch of abandoned highway. Jade leaned forward, scanning the map projected on her wristband.

"They moved her to the primary facility," she said. "Underground, west complex. Heavily shielded. We'll never reach it by air."

Kael's hands tightened on the wheel. "Then we go through the ground."

"Through?" Jade snorted. "Do you have any idea what's under that place? Bio-labs, drones, about two hundred men with guns, and a woman who likes scalpels."

He didn't answer.

"Look," she said, softer now. "We'll get her back. But we need the resistance. Arion's people have the tech to jam their scanners. Without them, we'll be walking into a tomb."

Kael's gaze stayed fixed on the road. "They think she's alone. They don't know what I am."

Lightning flashed across the horizon, illuminating the fury in his eyes—starlight, buried but alive.

Jade stared at him. "Oh, hell," she muttered. "You're waking up too."

Estelle

Back in her glass cell, Estelle lay still, counting the seconds between the pulses of the monitoring lights. Each one echoed through her body, and beneath that rhythm another beat answered—distant but familiar.

Kael.

Even across miles of earth and steel, she could feel him searching.

She pressed a trembling hand to her abdomen. "Hold on,"

she whispered to the life within her, to the man beyond the walls, to the star that refused to die.

"Hold on. He's coming."

Time bled inside the cell.

Hours, maybe days—no sunlight, no night, only the faint hum of machinery and the pulse of her heart trying to remember freedom.

The door hissed open with a rush of sterilized air.

Sori stepped in.

For a heartbeat, Estelle thought she was hallucinating again. Her friend's silhouette looked softer than memory—hair tied back, eyes red, trembling around the edges. She wasn't in a lab coat this time. Just plain gray fatigues with the AstraCorps insignia on her chest.

Estelle sat up slowly, her restraints clicking faintly.

"You shouldn't be here."

Sori swallowed hard. "They told me you were awake."

"After three months."

Sori's breath hitched. "I wanted to see you sooner. I swear I did."

Estelle's gaze hardened. "You turned me in."

The words hit Sori like a blade. Her lips parted, but no sound came at first. Then, quietly:

"I didn't want to. I had orders. You were my assignment, Estelle. I was—" her voice cracked— "I was your Tracker."

Estelle's pulse roared in her ears. "Tracker..."

"They pair us with Starborn," Sori said, eyes wet. "We live beside you, blend in, make sure you never remember who you are. Friends, lovers, mentors—it doesn't matter. Our job is to keep you safe from yourselves."

"Safe?" Estelle spat. "You mean contained."

Sori pressed her palms to the glass. "You have to understand—I didn't know what they were doing at first. When I met you, I thought you were just... a bright girl who loved the stars. Then I saw the files. The energy readings. I saw what you were. And the closer I got, the harder it became to report anything at all."

Her voice broke entirely now. "You became my family, Estelle. You— you reminded me what it was like to feel human again. And I hate myself for what I've done.

Estelle's throat burned. "Then help me."

Tears streaked down Sori's cheeks. "If I let you go... if you merge with—" she caught herself, glancing up toward the camera in the corner— "if you do what they fear, Earth won't survive. That's what they told us. Lyratheon's rebirth will drain this planet dry."

Estelle leaned forward, the cuffs clinking against the rails. "They lied to you, Sori. To all of you. Lyratheon doesn't take —it gives. Its energy sustains both worlds. It's not destruction, it's balance."

Sori shook her head like she wanted to believe but couldn't. "You don't know that."

"Yes," Estelle whispered. "I do. I've seen it in my dreams— Kael's memories, mine, everything the Keeper told us. Both worlds can coexist. They were always meant to."

For the first time, Sori's eyes softened with something like hope. "If that's true..."

Before she could finish, the door slid open.

A new shadow fell across the floor.

Director Voss stepped inside, his presence chilling the air like a storm. Two guards followed, silent and armed.

"Dr. Cho," *he said smoothly.* "You were not authorized for this visit."

Sori straightened instantly, wiping her face. "Sir, I was monitoring her vitals—"

"Leave us."

She hesitated. "Director—"

"Now."

Sori cast one last look at Estelle, whispering so faintly only Estelle could hear, "Hold on." Then she slipped out the door, leaving cold silence behind.

Voss approached the bed, his hands clasped behind his back.

"I see your conversation was... emotional," he said. "Touching, even."

Estelle glared at him. "You're monsters."

He smiled faintly. "Monsters are only gods seen from the wrong side of the glass."

He pulled a tablet from his pocket and tapped the screen. Estelle's medical data appeared—graphs, waveforms, and *an image that made her stomach drop.*

Her child's heartbeat.

"Your pregnancy is progressing faster than expected," he said calmly. "Half the gestational time of a normal human fetus. Remarkable. We estimate full cellular development within eight weeks."

Her voice shook. "You're not touching my child."

"Oh, Estelle..." Voss sighed, almost gently. "The baby won't reach full term. Once the labs extract viable tissue, the pregnancy will be terminated."

The words detonated in her mind. "No..."

"We'll harvest what we need—samples, amniotic fluid, perhaps a few organ cultures. Enough to replicate the starlight gene. The rest... will be handled humanely."

"Handled?" Her voice cracked, rising. "You mean killed!"

Voss's tone didn't change. "You misunderstand. It isn't life in the conventional sense. It's data. Energy. We simply wish to reproduce what nature has hidden from us."

Estelle lunged forward, restrained by the cuffs.

"You're insane!"

"Insane?" He smiled without warmth. "No. Visionary. In a few years, humanity will have its own celestial lineage—without the chaos of emotion, without the weakness of love. You'll be remembered as the prototype that worked."

Her pulse thundered. The monitors began to spike wildly, lights flashing.

"Calm yourself," Voss warned, but Estelle's voice came out like a roar.

"You can take my blood, my body, but you'll never take my light!"

For a split second, the air shimmered. The walls vibrated. The lights flickered violently, and one of the guards stumbled as sparks flew from the ceiling panels.

Voss stepped back, expression unreadable. "It seems your power is returning faster than anticipated."

He turned toward the door. "Double the sedation protocol. Memory wipe scheduled within seventy-two hours."

Estelle screamed as the guards approached. "You can erase my memories, but he'll find me! Kael will find me!"

Voss paused at the threshold, glancing over his shoulder.

"Let him try. We'll be waiting."

The door sealed behind him with a hiss.

Alone again, Estelle slumped forward, tears streaking her cheeks.

Her body trembled—not just from fear, but from the hum deep within her, pulsing stronger with every heartbeat.

She pressed a hand over her stomach, whispering through clenched teeth,

"I'll protect you. I'll protect both of us. Even if I have to burn this whole place to the ground."

The lights above flickered once more, and somewhere far away, under rain and lightning, Kael felt a pulse in his chest.

A frequency only he could hear.

CHAPTER 11
GENESIS

The Resistance Base

The entrance was hidden in plain sight – a rusted freight tunnel beneath the wreckage of an old coastal shipping yard. To the outside world, it was nothing but decay. But once Jade keyed in the access pattern and the metal doors groaned open, the air changed.

The tunnels thrummed with quiet life.

Lanterns powered by old solar grids lit narrow corridors carved into the stone. People moved with purpose – men and women with faint light under their skin, eyes that shimmered in the dark. Starborn. Almost all of them.

Kael followed Jade down the steps, his pulse pounding as he took in the sight.

"They came," he murmured.

Jade nodded. "Word spread fast. When Arion started...

reversing, it gave them hope."

They stepped into the central chamber – once a cargo bay, now transformed into a living command center. Screens glowed across the walls, maps drawn by hand pinned beside stolen satellite feeds.

And there – standing at the center, radiant and impossibly young – was Arion.

The last time Kael saw him, he'd been an old man in a rocking chair. Now, he looked no older than forty. His hair was streaked with silver, but his skin glowed with that same undercurrent of light that rippled through the Starborn veins.

When he saw Kael, he smiled. "You made it."

Kael stared, stunned. "You... you're–"

"Alive," Arion said simply. "And not done yet."

The other Starborns turned toward them, whispers filling the air. Dozens of them – maybe hundreds – their bodies faintly pulsing in rhythm. The energy was tangible, a hum of life that vibrated through the concrete walls.

Arion raised his hand, the room falling quiet.

“You all know why we’re here. AstraCorps has crossed a line. They’ve taken one of our own. They’re building something they call Project Genesis.”

He motioned to the monitor beside him. A projection flickered to life – diagrams, DNA strands, ultrasound scans that made Kael’s stomach twist.

“They’re using Estelle’s unborn child,” Arion said, voice grim, “as the template for their next generation of experiments. Cloned vessels. Manufactured fragments. They want to control creation itself.”

A murmur rippled through the crowd – outrage, disbelief, fear.

Jade slammed the datapad onto the table. It’s real, they’ve weaponized her blood.”

Kael’s jaw tightened.

Arion adjusted the scanner, frowning. “That’s not all. Look at this–”

A pulse flickered on the monitor. Small, rhythmic, steady.

“Energy signature?” Kael asked.

Arion nodded. “It’s faint, but distinct.. Hybrid frequency.

Celestial resonance spliced with human vitals. The child's heart beats at a harmonic rate – 9.7 hertz. It's almost... singing."

Kael leaned closer, the hum vibrating through his fingertips.

He didn't need the data to tell him. He felt it. A soft tug in his chest, as if someone small and far away had just breathed his name.

"Gene," he whispered as if their child had told him it's name.

The lights in the room flickered once, a soft acknowledgment.

Kael stepped forward, his voice sharp and shaking. "They think they can replicate what they don't understand. But they can't. Because what's inside her – what's inside us – isn't just energy. It's love. It's purpose. It's Lyratheon itself."

He looked around the room, meeting every glowing gaze. "They tried to bury our world. They shattered it. But even broken stars remember how to shine."

He took a breath, the anger boiling just beneath his ribs.

"For three months, they've kept her caged. For three months, they've tortured her and called it science. But look around you. Look at what's happening. Arion is living proof – when we stand together, when our frequencies align, we heal. The universe remembers us. Every one of you can feel it – the heartbeat of Lyratheon pulsing in your veins."

He raised his voice now, powerful, electric.

"They think we're a threat to humanity. But the truth is, we're its salvation. When the Starborn unite, we don't destroy – we create. We restore. We bring light where they've built only fear. So if they want a war, they'll get one – but not the kind they expect. We're going to bring Estelle home. We're going to bring our home home."

The chamber erupted in cheers and energy, the sound echoing like thunder through the stone. Some cried. Others lifted their hands, light spilling from their fingertips in shimmering waves that filled the air like starlight rain.

Jade stood to the side, watching him, chest rising and falling. For the first time, she saw not the fugitive pilot – but a leader.

Arion placed a hand on Kael's shoulder, his expression proud and heavy. "Your fire will lead them. But remember, young one – fire can both warm and burn."

Kael nodded. "Then let's burn the darkness."

Later That Night

The base had quieted. Most of the Starborn were asleep, their faint glow like constellations scattered through the tunnels.

Kael stood on a balcony carved into the rock, staring out at the moonlit waves beyond the cavern's mouth.

Jade found him there.

"You really meant every word, didn't you?" she said softly.

He didn't turn. "Every one."

She joined him at the railing, silence hanging between them. "You know, when I first found Arion, I thought it was a curse – this connection. This... pull. But watching you tonight, I finally understand what he meant. The heart cores – you and Estelle – you change the air around you. You make everything alive again."

Kael sighed. "That's what she does. Always has."

Jade looked at him then – long, searching. "She's lucky."

Before he could answer, she leaned in, her lips brushing his. It was soft at first, then desperate – a spark born of loneliness and borrowed hope.

Kael froze, then pulled back gently, shaking his head.

"Jade," he said quietly, "don't."

She stepped back, shame flickering in her eyes. "I'm sorry. I just—after everything—we've been together for months, and sometimes I forget..."

He placed a hand on her shoulder. "You didn't forget. You just felt the echo. The Starborn pull – it binds all of us when we're close. But what I have with Estelle isn't gravity. It's choice."

Jade swallowed hard, nodding. "You're right. Focus first. Save her. Rebuild after."

Kael turned back to the ocean. "No rebuilding without her."

They stood in silence, the only sound the distant hum of the sea – the pulse of something ancient stirring beneath it.

Far out on the horizon, faint lights danced across the

waves. Not *stars – not ships – something else.*

And Kael felt it.

Estelle's *pulse, faint but alive.*

He closed his eyes and whispered to the wind,

"Hold on, star girl. We're coming."

Estelle

They told her it would be painless.

That was the first lie.

The restraints bit into her wrists as the machines came alive. Electrodes *pulsed against her temples, cords coiled like veins across her arms, connecting her to the monitors that watched everything—her pulse, her dreams, even the rhythm of her thoughts.*

The air smelled like antiseptic and cold metal. The lights were too bright to be human.

Every time she tried to focus, the voice in her head—Kael's voice—faded a little more.

Remember me...

She clung to it, but the hum inside her was changing. The

starlight she carried began to dim, drained from her body in waves as the process began.

They called it Reprogramming.

But it felt like dying.

The Observation Room

Sori pressed a trembling hand to the glass, watching Estelle's body convulse under the current.

The monitors spiked—heart rate, energy output, resistance—before stabilizing again.

A technician made notes calmly beside her.

"Subject continues to show neural activity during sedation," he said. "It's like her mind refuses to go dark."

The hum drilled through her skull like static, separating thought from self. Obedience filled the spaces where her will used to live.

"Comply," the technician's voice commanded.

She wanted to. The compulsion was coded deep, written into her veins by the needles they'd used to make her forget.

But under the noise—

A whisper.

Estelle...

Kael's voice. Not heard, but felt. Like a vibration beneath her ribs.

The command repeated, louder now: "Comply."

Her lips parted. "I..."

Don't listen.

The whisper sliced through the static, raw and distant, but hers.

Her pulse spiked. The readings on the monitor stuttered.

Voss leaned toward the glass, watching as her eyes flickered open, their glow faint but defiant.

Somewhere inside the programming, a war had begun.

Sori couldn't speak. She could barely breathe. Every day for weeks, she'd stood here. Watching. Listening. Pretending not to care while inside her, something broke a little more each time Estelle screamed. She'd promised herself she wouldn't cry again. But when she saw the blood—golden red, shimmering like crushed starlight—flow from Estelle's *body into the containment vial, she turned away, tears burning her eyes.*

Inside the Lab

Director Voss entered without announcement. His presence always made the room colder. He walked with a stillness that unnerved everyone—like a man convinced the world already belonged to him. He stopped by the containment table, studying the vial of Estelle's blood under the lamp. It glittered faintly, alive even in stillness.

"Oh, you clever girl..." he murmured, holding it up to the light. "You hid your secrets in your veins."

He smirked and slipped the vial into his coat pocket. "Begin the transfusion."

The head scientist nodded. "Yes, sir."

The machines hummed louder. Tubes withdrew Estelle's luminous blood while another line pumped human blood in. The fluids mixed briefly in the tubes—light and dark, gold and red—then vanished into her body.

Estelle's back arched. Her mouth opened in a silent scream.

Monitors flared red.

"She's rejecting the input!" a tech shouted.

Voss didn't flinch. "Then increase the voltage."

Estelle thrashed against the restraints. Images tore through her mind—Kael's smile, the ocean, the stars. She tried to hold on to them, but each pulse of electricity stripped another memory away.

Her name.

His name.

The child.

All of it, unraveling thread by thread.

Observation Deck

Sori's nails dug into her palms until she bled. She couldn't watch anymore.

This wasn't containment. It was erasure.

Every day she'd told herself there was still time, that she could fix it, that she could save her. But now she knew if she didn't act soon, there would be nothing left to save.

She turned and walked away from the glass.

Her voice, when it came, was low and certain. "I'm breaking her out."

AstraCorps Command Briefing

Hours later, the boardroom buzzed with tension.

"Sir, the failed recruit from the Resistance has been recovered," an agent reported. "He talked—partially. Mentioned a base near the southern coast."

"Coordinates?" Voss asked.

"Unclear. Whatever tech they're using cloaks their frequency. We can't lock on."

Voss's expression hardened. "So they hide behind miracles now."

He turned to the head scientist. "Accelerate Genesis."

The man blinked. "Sir, the prototypes aren't viable. Cellular collapse begins within hours. Their bodies can't withstand the energy—"

"Then stop using human templates," Voss snapped.

He reached into his pocket and withdrew the vial of Estelle's blood, holding it up like a relic. It glowed faintly, as if aware.

"Use this."

The scientist hesitated. "Sir, that's her direct sample. It's too volatile—"

Voss's smile was thin. "Precisely why it will work. Merge it with the fetal cells. Speed up the development cycle. I want results before the week is out."

Two Days Later

Estelle drifted between pain and nothingness.

She no longer knew if the light she saw was real or memory.

When she could focus, she saw flashes—the lab, the nurses, Voss's silhouette in the doorway. Sometimes Sori's face, pale and weeping behind the glass.

Sometimes Kael's.

Always Kael's.

Hold on, star girl... we're coming...

She clung to the whisper, though she couldn't remember whose voice it was anymore.

The Genesis Lab

In the sterile glow of the sublevel facility, Voss *and his lead scientist stood before a sealed incubation chamber. Inside, something pulsed—a faint, rhythmic shimmer of light within translucent fluid.*

"Report," Voss *said.*

The scientist's voice shook. "Integration complete. We fused the fetal tissue with the Starborn DNA *sample. The other embryos failed as expected, but this one... this one stabilized."*

Voss *stepped closer, watching the small form within the glass.*

Tiny fingers twitched.

A *glow radiated from its chest like a heartbeat.*

A *slow smile crept across his face.*

"At *last," he whispered.* "A *child born of starlight and science. The first of our new creation."*

He placed a hand on the chamber's surface. "Welcome to the world, little one. Project Genesis... *lives."*

Far above the lab, in her dim cell, Estelle *stirred in her sleep.*

Her body was weak, her blood human, her mind fractured—but deep inside her, something answered.

A pulse.

A hum.

A flare of starlight that refused to die.

The *machines spiked.*

Monitors *screamed.*

And *for the first time in months, the cameras captured something impossible...a faint glow reappearing under* Estelle's *skin.*

CHAPTER 12
THE SPARK

The Escape

The facility was louder than usual that night.

Sirens blared faintly down the corridors, though no one in the upper levels seemed to know why. Sori moved quickly through the sterile halls, her head down, badge clutched tight in one shaking hand.

She'd spent weeks planning this – memorizing guard rotations, learning the power grid layout, and rerouting the security feeds to loop thirty seconds of static.

Thirty seconds. That was all she'd get.

Inside her chest, her heart pounded like it wanted out.

She stopped outside Estelle's door.

The glass was fogged from the humidity inside, the faint light of monitors flickering against the walls.

Estelle lay on the cot – pale, thin, eyes half-open but

unfocused.

"Hey..." Sori whispered as she slipped inside. "It's me."

No response.

She knelt by the bed, gently touching her arm. "Estelle? It's over. I'm getting you out."

Estelle blinked slowly. Her gaze wandered, unfixed, as if she were seeing through fog. "Who...?"

Sori's breath caught. "It's Sori. You remember me, don't you?"

A pause. Then, faintly: "Sori..."

Relief almost knocked Sori over. "Yeah, that's it. I'm here. We're leaving."

Estelle's eyes shifted. Her hands—weak, trembling—moved to her belly. She pressed down, as though searching for something that wasn't there.

"There's nothing," she whispered. "Why—why is there nothing?"

Sori froze. "Estelle..."

"What happened?" Estelle's voice was hoarse, rising. "I—something was here."

Sori's throat closed. "You were pregnant," she whispered. "They... took it. They told you you weren't ready, and they —"

"No." Estelle's tone sharpened, her eyes clearing slightly. "You're lying."

Sori shook her head, tears spilling freely now. "I wish I was. They harvested your fetus for Project Genesis. They made—" she stopped herself, choking on the words. "They made a child out of you. Out of it."

Estelle's breath hitched. She looked down at her trembling hands as though seeing them for the first time. "My blood..." she murmured. "It was mine."

Then Sori said the name.

"Kael is coming for you."

The name hung in the air like a bell—sharp, metallic, hollow.

Estelle blinked, tilting her head slightly. "Kael?"

"Yes," Sori said softly. "The man you love."

Estelle's expression didn't change. "I don't know him."

Sori's heart cracked in two. "He's your other half, Estelle.

Your Starborn twin—"

"I said I don't know him!" Estelle *snapped, the monitors around them flaring for a moment, lights pulsing like lightning.*

Sori flinched but didn't move away. "Then I'll remind you. I'll make it right. I swear I will."

But Estelle*'s eyes were different now—cold, hollow, stripped of their glow. What was left inside wasn't light.*

It was rage.

"Then start by getting me out of this hell," she said quietly.

Sori

The alarms hit just as she wheeled Estelle *into the main corridor. Red lights bathed the walls, and a voice echoed over the intercom:*

"Containment breach in Sector 3B. *All personnel to lockdown stations."*

Perfect. Her distraction had worked.

She guided Estelle *quickly down the hall, avoiding cameras as best she could. Her security codes would buy a few* minutes before AstraCorps realized she wasn't returning to her post.

Estelle stumbled once, weak from the blood loss, and Sori caught her.

"Almost there," she whispered.

They passed through the surgical ward, the smell of antiseptic sharp in the air. Monitors still glowed behind locked glass, showing charts and images of embryos—rows of small forms suspended in fluid.

Estelle stopped, staring. "What is this?"

Sori hesitated. "The Genesis Lab."

Estelle's eyes flickered. For a second, her pupils flashed gold, memory trying to claw its way back.

Then her voice came low, trembling: "What did they do to me?"

Sori pushed her forward, her tone urgent. "I'll explain later. We have to move."

They burst through the loading bay doors, alarms howling now from every floor. Sori had hidden an old AstraCorps vehicle here weeks ago—a dusty transport car she'd rigged *to bypass the ID scanners. She helped Estelle inside, slamming the door.*

"Hold on."

The tires screeched, rubber tearing against the concrete as they raced down the back access road.

Sori didn't look back. She didn't need to. The fire alarms, the flashing lights—they were enough.

She'd done it.

She'd broken her out.

But when she looked over at Estelle in the passenger seat, the girl she'd once known wasn't there anymore.

Estelle's gaze was fixed ahead, her face illuminated by the red light of the dashboard. There was no fear in her expression—only quiet, simmering wrath.

"They took everything from me," she whispered. "Everything."

Sori swallowed hard. "We'll get it back. I promise."

Estelle turned her head slowly, eyes glinting faintly. "You'll help me kill them all?"

Sori's breath faltered. "Estelle..."

I'm done being a subject," Estelle said. "Now they'll learn what a Starborn really is."

The Assault

Kael

The convoy cut through the night like a blade.

The Resistance had never moved this fast before – not even when AstraCorps raided the first sanctuary. But this time, Kael was leading, and there was no hesitation left in him.

"Teams one and two breach through the west hangar," Jade said through the comms. "Teams three and four, hold perimeter and jam the external frequencies. We've got thirty minutes before reinforcements arrive."

Kael's grip tightened on the controls of the lead vehicle.

"We won't need thirty."

Arion's younger voice came through the radio, calm but urgent. "Remember what's at stake. AstraCorps has already accelerated Project Genesis. If they've managed a viable Starborn clone, that child could become a weapon – or worse"

Kael's eyes burned against the dashboard glow. "They won't use her. Or her child."

He looked out at the horizon – the black silhouette of AstraCorps' facility gleaming under floodlights. Somewhere inside that fortress was Estelle. The woman who had died, been reborn, and now, was suffering because of him.

The thought carved through his chest.

"Ready?" Jade's voice snapped him back.

Kael exhaled. "Let's end this."

The convoy hit the compound like thunder.

Explosions ripped through the fences, energy charges flashing white and blue. Drones dropped from the sky in bursts of flame as electromagnetic pulses fried their sensors.

Kael leapt from the vehicle before it fully stopped, plasma rifle in hand, moving like instinct – all precision and purpose. Starborn light glowed faintly in his veins, burning brighter with every heartbeat.

Inside, alarms screamed. Guards poured in from both corridors, but the Resistance fought with something

AstraCorps couldn't understand – unity. The air shimmered with their combined energy, the pulse of Lyratheon itself rippling through the halls.

Jade shouted from behind cover. "We've got two minutes to hit the main lab!"

Kael nodded, eyes locked ahead. "Then we go through."

He slammed his hand against the reinforced door. The metal shuddered – once, twice – then exploded inward, torn apart by raw energy.

The team stormed in.

And froze.

Rows of tanks lined the lab, filled with a luminescent fluid that bathed the room in eerie blue light. Inside them floated small, motionless forms – not quite human, not quite light.

"God..." one of the rebels whispered.

Jade's hand trembled as she scanned the data on the console. "All failed subjects. But... there's one active chamber."

Kael followed her gaze. In the center of the room, a single pod pulsed softly – the glow steady, rhythmic, alive.

He approached slowly, his boots echoing against the tile. Inside the chamber floated a child – no more than a few weeks old, tiny, perfect. Its chest rose and fell, golden light flickering beneath its skin.

Kael's knees weakened. "Estelle's child...our child"

Arion's voice came through the comms, hushed with awe. "The Genesis success."

Jade swallowed hard. "She wasn't lying. They actually did it."

Kael's jaw tightened. "Where's Estelle?"

Jade ran another scan, her brow furrowing. "There's... nothing. Her cell's empty. No trace."

"What do you mean empty?"

"I mean gone. She's not here."

Kael slammed a fist against the console, the impact sparking electricity. "They moved her!"

Arion's voice came through again, strained. "No... I think someone moved her. I'm reading a rogue vehicle departing the south gate ten minutes ago. AstraCorps didn't authorize it."

Kael froze. "Sori."

Jade looked at him. "You think she—"

"She did. She got her out."

For a moment, relief flickered in his chest – then vanished as the reality hit him. If they were running, they were running alone.

He turned back toward the child. The glow inside the tank pulsed faster, reacting to his proximity.

"Kael..." Jade said quietly. "What do we do?"

He stared at the child – the living fusion of Estelle's body, her blood, her light. A piece of her...them that shouldn't exist, but somehow did.

His voice was low, steady. "We take it. I can't leave our son here to become their weapon."

Jade hesitated. "Son?"

Kael nodded once. "I can feel him."

They disconnected the power supply, sealing the pod in a containment shell. The alarms screamed louder now – security had breached the lower levels. Arion's voice echoed through comms, urgent. "Kael, you have incoming from the east corridor. Forty seconds!"

Kael lifted the containment shell in his arms. "We're moving!"

The Resistance fought their way back through the halls, energy blasts flashing in every direction. Starborn light flared from Kael's hands, disarming drones mid-air, shattering armored suits like glass.

By the time they reached the exit, smoke filled the corridors and the building burned behind them.

Jade slammed the door to the transport closed, coughing. "You've got him?"

Kael looked down at the containment pod in his lap. The baby inside glowed faintly – tiny fingers curling as though sensing him.

"Yeah," he said softly. "I've got him."

Arion's voice crackled over the radio. "Then go. Get back to base. We'll cover you."

Kael's jaw set. "Not yet."

"Kael–"

"I'm not leaving without her."

The vehicle roared to life, tires screaming as it tore down the highway.

Meanwhile

On a lonely stretch of backroad miles from AstraCorps, Sori's car sped through the dark.

Estelle sat in the passenger seat, silent, her eyes reflecting the lights of passing cities. The wind whipped through the cracked window.

Sori glanced at her, voice trembling. "You don't remember him, but he's real. He's fighting for you. You have to believe that."

Estelle didn't respond. She just stared ahead, her hand pressed flat against the glass, watching the sky.

Somewhere above, the stars flickered – faint, pulsing in rhythm.

When she finally spoke, her voice was soft but certain.

"I don't remember him," she said. "But something tells me he remembers me."

She turned her head slightly, her eyes beginning to glow –

faint, gold, alive.

"And when he finds us... they'll wish he hadn't."

AstraCorps

Back at the burning facility, a single monitor still flickered in the wreckage.

A message blinked red on the screen:

PROJECT GENESIS: SUCCESS.

SUBJECT TWO – ACTIVATED.

CHAPTER 13
WHEN THE STARS WENT SILENT

A World Without Stars

No one remembered the exact moment the light went out. It didn't happen with a sound or a flash—no explosion, no global alert. One night, the sky simply stopped shining.

At first, people thought it was dust in the upper atmosphere or a satellite interference. Astronomers scrambled for answers, but every telescope returned the same thing—nothing. No stars. No glimmer. No trace of what once was.

The moon still hung in the heavens, dull and colorless, but even it seemed bruised, as though ashamed to be seen.

Cities dimmed their skylines to conserve power. Crops began to wilt. Temperatures dropped by degrees that shouldn't have been possible. The world grew colder, quieter, like it was holding its breath.

Scientists called it the Veil Phenomenon.

But in the streets, people whispered something older:

"The sky's punishing us."

Washington D.C. – NASA Headquarters

Every major network streamed the emergency press conference. Rows of reporters filled the auditorium, murmuring beneath the flicker of failing lights.

Dr. William Hyder, NASA's chairman, stepped up to the podium, his silver tie glinting like metal under the artificial lights. He looked tired—drawn—but his voice was steady.

"Good evening. We understand the public's concern regarding the recent atmospheric anomaly. As of 0300 hours yesterday, global optical sensors recorded the disappearance of the last visible starlight within the night sky."

Cameras clicked.

"We have confirmed this is not the result of an astronomical event such as a supernova, magnetic storm,

or cosmic interference." He paused, choosing his next words carefully. *"Our current working theory suggests a temporary shift in the Earth's axial resonance frequency—a rare but not unprecedented fluctuation in the planet's rotational energy field. Essentially..."* He smiled, rehearsed. *"The Earth has blinked."*

A *murmur rippled through the audience.*

He continued, *"We anticipate the stars will reappear once equilibrium is restored.* NASA, *in partnership with the* Department *of* Defense *and private contractors, are monitoring the situation closely."*

A *reporter shouted from the crowd,* "Is AstraCorps *involved?"*

Hyder's *jaw tightened.* "AstraCorps *is a private organization. Their research does not fall under* NASA *jurisdiction."*

Another *voice: "But weren't they working on star-energy replication—?"*

Hyder *cut them off smoothly. "As* I *said, no connection has been confirmed. We advise calm. The stars will return.*

They always do."

The Run

Rain hammered the cracked windshield as Sori's old car cut through the backroads of Tennessee.

The night sky above was blacker than pitch—no moon, no stars, just an oppressive void pressing down from every direction. Estelle sat slumped in the passenger seat, head resting against the glass. Her breath fogged the window in soft bursts, each one glowing faintly gold before fading.

She had been quiet for days.

Too quiet.

Sori risked a glance at her. The woman looked hollow—like a reflection of who she once was. But sometimes, when lightning flashed, she caught glimpses of the other Estelle: the one whose eyes burned with galaxies.

That was the part that terrified her most.

Because those eyes didn't always belong to Estelle anymore.

They had been hiding in abandoned barns, motels, anywhere the scanners wouldn't pick up heat signatures.

Sori stayed awake through most nights, researching on stolen tablets, trying to decode the signal patterns she'd copied from AstraCorps before they escaped.

The data all pointed to the same thing—something had been embedded inside Estelle. Not mechanical. Not even biological. Something worse.

A control sequence... A living failsafe.

Every Starborn carried energy, but only Estelle's had been tuned to a specific frequency—a leash woven into her DNA. Whenever her emotional spikes reached a certain point, it triggered external activation signals.

Sori didn't know exactly what that meant, but she'd seen it once before.

The nightmare.

That night, they'd been camped behind an abandoned gas station on the outskirts of Baton Rouge. The fire burned low, wind whipping through broken signage.

Estelle had begun to twitch in her sleep, mumbling incoherently. Then she screamed. Sori jumped up, rushing to her side, but froze when she saw her eyes.

They weren't human anymore.

Her irises fractured into shards of light, spinning like broken mirrors. Code—actual, visualized binary code—flickered across her pupils, projected like a digital overlay. Her skin pulsed with circuits that weren't metal but light itself.

"Estelle," Sori *whispered.* "Wake *up.*"

Estelle's *voice came out strange, layered.*

"We *are online.*"

And then the air around them shimmered, vibrating like a struck chord. The ground cracked beneath her, and the temperature dropped ten degrees in seconds.

Sori *panicked, shouting her name again, and in a blink, it stopped.*

Estelle *gasped awake, drenched in sweat, eyes glassy.*

"What *happened?" she whispered, trembling.*

Sori *couldn't answer.*

Because she was too busy hiding the terror in her own eyes.

Now, *days later,* Sori *drove through the endless dark,*

replaying the moment again and again in her head.

Estelle stirred beside her. "Where are we going?"

"South," Sori said quietly. "There's a safehouse near the gulf. Off-grid."

Estelle didn't answer, just stared out into the void.

After a long silence, she said, "The stars are gone."

Sori's grip tightened on the wheel. "Yeah."

"Do you think it's my fault?"

Sori glanced at her, voice trembling. "No. It's theirs. They broke you, Estelle. They broke the sky trying to play God."

Estelle turned, eyes half-lit with faint gold. "Then maybe God finally noticed."

Late That Night

The safehouse was a weather-beaten cottage deep in the bayou, surrounded by swamp mist and cicada song.

Sori worked by candlelight, hunched over her laptop, running diagnostics on Estelle's vitals through the patchwork monitor she'd built.

Lines of alien code flickered across the screen. Each spike matched Estelle's brain waves during the nightmares.

"It's in your blood," Sori whispered. "They rewrote you."

She closed her eyes, exhaustion finally cracking her voice.

"I'm going to fix this. I don't care how. I'll find it, pull it out, burn it—whatever it takes."

Behind her, Estelle lay still on the cot. But her eyes were open, staring at the ceiling, and for the briefest moment, Sori saw the same flicker of light—binary code across her pupils.

Estelle blinked once. Twice. Then it vanished.

Her voice, when it came, was calm but cold.

"Then hurry," she said. "Because I think they can see me again."

The Living Key

The resistance's new refuge lay beneath a gutted hydro-plant, its turbines long silent, its chambers thrumming instead with the pulse of gathered Starborn. Water dripped from the ceiling, echoing like a slow heartbeat.

Kael set the containment pod on the table in the center of the war room. The child inside glowed faintly—each breath a ripple of soft gold that filled the chamber with warmth the world above had forgotten.

Arion, now firmly in his prime, approached. Lines of age were gone, replaced by a steady radiance that made the others instinctively fall silent.

"This," he said, resting a hand above the pod, "is why the sky darkened."

Jade frowned. "You're saying one baby did all that?"

Arion shook his head. "Not the child—what he represents. When AstraCorps used Estelle's blood and the fragment of her unborn child, they awakened a tether. Lyratheon is reaching for him, and the Earth is resisting. Two worlds locked in the same breath."

Kael stared at the sleeping infant. "What happens if the tether snaps?"

"Then both worlds collapse inward," Arion replied quietly. "But if the link is completed—if Estelle and the child reunite—Lyratheon's core reignites. The stars return."

Kael's hands tightened on the edge of the table. "Then we

find her before they use him."

The others murmured assent, but Arion held up a hand. "There's more. The boy isn't just a fragment. He's a key. His frequency bridges the old and new. Through him, the gate to Lyratheon can be reopened."

Jade's eyes widened. "You mean—home."

"Home," Arion confirmed. "Or destruction. It depends on who controls the resonance."

Kael turned away, chest tight. "He's her child. Our child. He belongs to us, not them."

Arion's gaze softened. "Then you must protect him, even from yourself. He's bound to you by light and blood, but his will is his own. When the time comes, the child will decide which world survives."

Kael looked down again at the pod. The baby's tiny hand pressed against the glass, almost as if reaching for him. Kael placed his palm against it, a rush of warmth shooting up his arm.

"I swear," he whispered, "I'll bring her back. And when I do, we'll bring the stars home together."

Arion nodded. "Then may the universe remember your promise."

The Call

The hydro-plant base was never quiet anymore. The hum of Starborn energy filled the air at all times – a low, living vibration that seemed to echo the baby's pulse in the containment pod.

Jade stood beside Kael, watching as Arion and the other resistance engineers studied the readings projected on the chamber wall.

"Every hour it grows stronger," one of them said, pointing to the hologram. "We're seeing measurable increases in atmospheric distortion, especially at night. The light that should reflect off the upper atmosphere—it's being drawn inward. Consumed."

Arion folded his arms. "By him."

Kael turned sharply. "The baby?"

Arion nodded. "He's absorbing the starlight. Instinctively. The fragments of Lyratheon scattered through the cosmos are answering his frequency. He is... collecting them."

"So he's killing the sky?" Jade said, disbelief tinging her tone.

"No," Arion replied quietly. "He's calling it home. But the Earth cannot sustain that much resonance. The balance is breaking faster than expected."

Kael looked down at the pod. The baby's chest glowed faintly, each pulse brighter than the last. "Then we need Estelle. She's the other half. Without her, he doesn't know when to stop."

Before Arion could answer, one of the technicians called out, "Incoming transmission. Private line."

Jade frowned. "From who?"

The tech hesitated. "It's... encrypted with AstraCorps coding."

Kael's eyes narrowed. "Put it through."

The screen flickered to life, static hissing before resolving into a grainy image of Sori. She looked exhausted—hair tangled, eyes rimmed red—but alive.

"Kael," she said, her voice hoarse. "If you're hearing this, don't bring anyone else. Just you."

Jade's jaw clenched. "That sounds like a trap."

Kael shook his head. "That's her. She's alone."

On the screen, Sori glanced over her shoulder, clearly in hiding. "I have Estelle," she continued. "But she's... not herself. You can't let her near the others. She doesn't have free will anymore."

The room went still.

"What do you mean?" Kael asked quietly.

Sori looked directly into the camera, tears glinting in her eyes. "They turned her into something else. A conduit. A weapon. AstraCorps embedded a control frequency into her DNA. It activates through emotion—through memory. If she feels too much... they can take her."

Kael's heart sank. "You're saying they can control her remotely?"

Sori nodded. "She doesn't know. But I've seen it. When she dreams, her eyes digitize. Her pulse syncs with AstraCorps signals. I think they can see what she sees. I think she's their satellite."

Arion's face paled. "If that's true, then she's a walking beacon. They'll find her—and us—if we bring her here."

Sori's voice cracked. "That's why I can't bring her to you. Not yet. I'm trying to figure out how to sever the link. I'll send coordinates when it's safe."

Kael leaned closer to the screen. "Where are you?"

She hesitated. "You don't want to know. The less you do, the longer I stay alive."

The feed crackled. The static deepened.

"Kael," she said softly, "when you see her again... she won't remember you. Don't take it personally. She's still in there somewhere."

Kael's throat tightened. "Sori, listen to me—"

The feed cut out.

For a long moment, no one moved. The silence was unbearable.

Finally, Jade exhaled. "She's right about one thing. Bringing Estelle here could expose us. But if that baby keeps pulling in the light, the whole planet will freeze before she finds a cure."

Kael turned toward Arion, fury and heartbreak warring in his voice. "So what are you saying? We do nothing?"

Arion met his gaze evenly. "I'm saying that every action now has cosmic consequence. If the child is absorbing starlight, he's rewriting gravity itself. Estelle's link to him is the only stabilizer—but if AstraCorps still controls her, connecting them could destroy both worlds."

Kael shook his head. "Then we find a way to break the link first."

Jade placed a hand on his arm. "She asked for you alone. You think that's wise?"

Kael's voice came low, steady. "She saved her once. Maybe she can again."

He looked toward the dark horizon beyond the base tunnels, where the faint glow of the pod pulsed behind him like a heartbeat.

"I'll go," he said. "If there's even a chance she's still in there, I have to try."

Arion nodded slowly. "Then go quickly. Because the longer she's out there, the closer AstraCorps comes to finding her."

Kael turned toward the exit, the shadows swallowing him.

Behind him, the baby stirred.

And above them, for the first time since the raid, a faint star flickered into view then blinked out again.

Found

Far away, in a motel off an abandoned interstate, Sori sat on the edge of the bed, watching Estelle sleep.

Lightning flared outside, illuminating her face.

The veins beneath her skin pulsed faintly with light – code-like threads weaving through her neck, her temples.

Her body was quiet now, but Sori could feel the hum in the air, faint but rising.

She whispered to herself,

"Please, Kael. Don't take too long."

Estelle stirred, murmuring something in her sleep – a name that came out like static.

Then the lights in the room flickered. The air grew heavy.

Sori turned toward the window. In the distance, a single

black drone drifted down from the clouds, its sensors glowing red.

Her stomach dropped.

"They've found us," she *whispered.*

CHAPTER 14
THE LOST LIGHT

The night the drone found them, the world was unnervingly quiet.

No stars. No wind.

Only the hum of the machine hovering outside the safehouse, a single red light pulsing through the rain like a heartbeat.

Sori instincts screamed before her mind could form the words. She grabbed Estelle's wrist.

"Run."

But it was already too late.

The air outside fractured into static – invisible frequencies rippling like heatwaves – and before they reached the door, a shockwave slammed them both backward.

Estelle hit the ground hard, her vision stuttering between darkness and light.

She heard Sori shouting her name, boots pounding the floor, the whine of energy rifles charging.

Then came the voice over the loudspeaker:

"Subject Auryn acquired. Handler compromised."

The rest was silence.

The Reconditioning

When Estelle woke again, she was back inside AstraCorps.

The walls gleamed white.

The air was sharp with ozone and disinfectant.

She didn't move right away.

She just lay there, staring at the ceiling, counting the faint flickers of light that occasionally ran through it like veins.

A voice cut through the static of her thoughts.

"Welcome home, Estelle."

She turned her head slowly. Director Voss stood at the foot of the bed, hands clasped behind his back, immaculate as ever. His eyes shone with sterile delight.

"You gave us quite the chase," he said. "But you didn't think you could hide forever, did you? We made you, after all."

Estelle's mouth was dry. "You broke me."

Voss smiled faintly. "We refined you."

He gestured, and two guards dragged Sori into the room.

She was bruised, restrained, blood streaking her cheek.

Estelle sat up instantly, the restraints around her wrists tightening as she pulled against them. "Don't touch her!"

Voss tilted his head. "Oh, but she's already touched you more than she should have. The loyal handler turned sympathizer. Did you really think we wouldn't find out she was feeding intel to the Resistance?"

Sori met Estelle's gaze, eyes wide with regret. "I'm so sorry Este-"

The sound of the gunshot cut her off.

Estelle screamed.

The restraints tore into her skin, the light in her veins surging bright gold before dimming again as the neural inhibitors kicked in.

Voss sighed. "Such a shame. She was one of our best."

He turned back to Estelle, voice calm and clinical.

"Now. Let's talk about your child."

The Weaponization

They strapped her to the interface chair again, wires threading into her spine, electrodes pulsing against her temples.

"Don't fight it," Voss *said gently, almost like a father. "We've optimized your design. You'll be able to access every Starborn frequency on the planet now. Imagine it – you, the bridge between worlds."*

Estelle's voice was a rasp. "I don't want this."

He smiled. "You don't get to want anymore."

The machine activated.

Pain tore through her body in luminous arcs, her vision flooding with maps – energy signatures, heat pulses, starlight readings. Every flicker of life across the planet painted itself in her mind like a constellation.

And among them... she saw him.

Kael.

A golden signature burning brighter than the rest, moving toward something—someone.

Her breath caught.

The machine whined louder as her pulse spiked.

Voss's eyes lit up. "Yes... there he is. The other half of the twin core."

Estelle shook her head violently, tears streaking her face. "No—stop—"

But the code within her obeyed.

Her heartbeat synced to Kael's.

Her mind stretched across distance, and suddenly she could feel him – the heat of his body, the echo of his pain, the memory of his voice.

For a fraction of a second, the programming faltered.

A whisper escaped her lips.

"Kael..."

Then her eyes went dark again – pupils dissolving into fractal light.

Voss leaned in. "Good girl. Now take us to them."

They dragged Estelle back in chains, but this time the fear didn't take her—the visions did. Somewhere between the blindfold and the cold metal floor of the transport, the world fractured. Light spilled across her mind like a cracked star, voices layering over each other—warnings, directions, memories that weren't hers. She felt herself drifting, floating, pulled through a current she didn't understand. Every heartbeat echoed with a place she had never seen but somehow knew.

By the time the blindfold was ripped away, something inside her had changed. Something the soldiers saw instantly. Her eyes—no longer just Estelle's—glinted with that faint, impossible shimmer. It was the Starborn awakening. Estelle was gone, AstraCorp's control held on to her too tight for her to break free. So they stopped treating her like a prisoner and started treating her like a compass.

Days blurred as they studied her reactions—where her pulse quickened, where her skin prickled with that strange magnetic pull. Until finally, the commander ordered it: "Put her in front. Wherever she's leading us... that's where the Starborn will be."

And Estelle, still shackled, still burning with questions she didn't dare ask, stepped out into the harsh daylight at the head of a roaring convoy. Engines thundered behind her. The horizon trembled with possibility. She didn't know what waited for her out there. But the pull was stronger now....Hungrier...Certain. And with every step, she felt it guiding her straight toward the destiny she had never wanted—yet could no longer outrun.

Kael's Preparation

Deep beneath the ruined hydro-plant, Kael felt it before he saw it – a sharp, electric pain behind his eyes.

Like someone had reached into his chest and twisted.

Arion turned sharply. "You felt it too."

Kael nodded. "She's alive. And... close."

Jade frowned. "Close? You sure it's not another Starborn?"

"No." Kael's voice was low. "I'd know her anywhere."

He walked to the holographic map glowing across the wall, pointing to a location pulsing faintly on the perimeter.

"There. That's where they're coming from."

Arion stepped forward, brow furrowed. "You think she's leading them here?"

Kael didn't answer. He didn't have to.

The look in his eyes said everything.

The Moment

As AstraCorps' assault vehicles rolled toward the valley, Estelle stood at the front of the convoy, silent, her eyes burning gold under the night's black sky.

She felt nothing.

No fear.

No sorrow.

Only purpose.

But deep beneath the programming, something human stirred. A flash of warmth. A voice echoing through static.

"Always."

It was faint, like a forgotten melody – but it was enough to make her hesitate. The command frequency in her mind screamed for obedience, pulling her toward the target that pulsed like a heartbeat through the network.

Kael...Her mission...Her love.

She whispered to the void, "I'm coming home."

The Storm

The first explosions hit before dawn.

The valley lit up in violent flashes — streaks of red and white slashing through the fog like the birth of new stars.

Kael was already in motion before the alarms finished sounding. "Shields up! Evacuate non-combatants to the lower tunnels!"

Jade fell in beside him, weapons drawn. "How the hell did they find us?"

Arion didn't answer. He didn't need to. The same realization was already dawning in Kael's eyes.

Estelle.

She was out there.

He could feel her — her pulse in his own chest, a frequency only he could hear.

The Arrival

On the ridge above the compound, AstraCorps forces lined up like shadows against the dim horizon.

At their center stood Estelle Auryn, her hair whipping in

the wind, her golden eyes reflecting the lightning that split the sky. To the soldiers behind her, she was a miracle of science – the perfected hybrid. To Voss, she was the key to conquering creation itself. But to herself, she was nothing. The commands in her mind came like whispers layered over her heartbeat.

Target: Resistance Nexus.

Objective: Eliminate all Starborn signatures.

Retrieve Child Unit G-1 (codename "Gene").

Her jaw clenched. Her voice came out low, almost reverent. "Execute."

The world ignited.

The Resistance fought hard, but the first wave tore through them. AstraCorps drones rained down from the sky, releasing bursts of sonic energy that cracked rock and bone alike. Kael ducked behind a barrier, returning fire, his starlight pulse blasting through an armored unit. "They're pushing straight through—why?!"

Jade shouted over the chaos, "Because they're following her!"

He turned—and time stopped. Estelle moved through the smoke like a ghost wrapped in gold. Every step shimmered with refracted light. Bullets curved away from her as if afraid to touch her.

"Estelle..." he whispered.

The others saw only a weapon. He saw the woman who used to trace constellations on his skin.

"Hold your fire!" he barked.

But the others didn't listen. Shots rang out. The air erupted in light. Estelle raised her hands reflexively – and the light obeyed. A wave of pure energy rippled outward, flattening everything in a fifty-yard radius.

When the dust settled, half the valley was ash and silence. Kael was on his knees. Around him, some of the Resistance lay dead – Starborn who had followed him for months, turned to scorched glass.

Estelle blinked slowly, the glow fading from her skin as the override subroutines kicked in again. Her expression was blank, her voice metallic.

"Target area neutralized. Confirm next objective."

Kael rose unsteadily. "You're not their property. You're Estelle. Look at me."

She didn't move. He walked toward her slowly, ignoring Jade's warning yell behind him.

"Estelle, it's me."

Her head tilted slightly, recognition flickering somewhere deep behind those golden, digital eyes.

"Kael..." she said softly.

Then her voice distorted. "Kael Verran. Priority One Target. Terminate."

She lunged.

Kael barely deflected the strike – her strength wasn't human anymore. Every blow she landed crackled with starlight, each hit a small supernova. He blocked, dodged, desperate not to hurt her.

"Fight it!" he shouted, grabbing her wrists. "You're stronger than them!"

Her eyes blazed. "I am proprty of AstraCorps."

"No." He pulled her closer, forehead to forehead, their light flaring as one. "You're mine."

For a heartbeat, her eyes cleared – and she gasped like someone surfacing from drowning.

"Kael...?"

Then the override reasserted. Her hand shot forward, driving a blade of condensed light into his side. Kael stumbled, blood spilling between his fingers – not red, but faintly glowing gold. Estelle froze, horror flashing through her features as fragments of her own humanity bled through the code.

"Kael, I didn't–"

He coughed, reaching for her cheek. "You didn't. They did."

Her breath hitched. Something inside her cracked, pain and memory colliding like twin suns.

He smiled faintly through the pain as blood spilled from his lips. "Do you remember what I told you once? About forever?"

Her lip trembled. "Stop..."

He leaned closer, and kissed her lips before whispering against her ear.

"Always."

Even though the word hit her like a current, she still raised her weapon to strike again. He caught her wrist mid-swing, fingers digging into her pulse. The world slowed. In that stillness there was a spark. Not from the weapons or the energy storm around them, but from touch. A warmth so familiar it hurt. Her mind flashed – not to the lab, not to pain – but to another life.

Hands brushing on a sunlit field.

A laugh.

A promise.

Always.

Her pupils dilated, her body jerking as light flooded every vein. She screamed, collapsing to her knees as the control circuitry burned out in a burst of golden fire.

The drones above exploded mid-air. AstraCorps comms went silent.

And when she looked up again, her eyes were her own.

"Kael..." she sobbed, catching him as he fell.

He smiled – weak, blood at the corner of his lip, eyes burning with relief and something deeper.

"I knew you'd come back to me," he whispered.

A cough shook through him, but he didn't look away. "No amount of science... no rewiring or walls between us..."

His voice broke, soft but sure. "None of it could erase what we are." He reached for her hand, his thumb tracing the lines of her palm as if to remind himself she was real. "I told you I'd find you," he said, breath unsteady, "And I did."

Behind them, the battlefield burned – a sea of light and ruin, engulfing the remaining AstraCorps soldiers.

Jade and the surviving rebels reached them minutes later, hauling Kael toward the evac transport. Estelle followed, her face streaked with tears and ash. No one spoke. The silence was sacred.

When the transport lifted into the night, the Earth below was still black, starless – but for the first time, faint glimmers began to pierce through the clouds. One by one, small points of light blinked back into the sky.

Not enough to restore the constellations, but enough to give hope.

CHAPTER 15
THE YEARS BETWEEN STARS

The world didn't end in fire that night – it ended in silence.

When the alarms faded and the smoke cleared, there was nothing left of AstraCorps but ghosts and data dust.

For weeks, Estelle drifted between waking and sleep, her mind a patchwork of flashes – Kael's voice, Sori's tears, the hum of her own pulse under foreign stars.

She didn't remember the escape.

Only that she woke in the backseat of an old truck, the sky stretched endless above her, and Kael's hand was in hers.

They didn't speak for a long time. They didn't need to.

They hid first.

A month in the desert. Another in the ruins of an abandoned observatory. Kael found food, Estelle found quiet – and in the space between, they found each other again.

Her memories came back in pieces, fragile but alive.

Each one a constellation reassembling itself in her mind – flashes of laughter, of starlight, of home.

Estelle looked down at the baby who cooed and curled his tiny fingers around hers.

He was hers. Theirs.

The moment she held him, everyone in the room felt it – a quiet shift, like the universe had just found its balance. For the first time since Lyratheon, Estelle felt whole.

Kael smiled, watching them, but a flicker of jealousy tightened in his chest. He forced it down, reminding himself that Estelle had never been a mother in any of their past lives. This was new for her – precious, fragile – and he was terrified it would change everything.

She didn't understand, not yet, how much the others had sacrificed for this moment, how long he, Jade, and Arion had fought to bring their people together. To her, this child was the miracle.

To them, he was also the mission.

But when he saw her cradle the boy, saw the light return to her eyes, Kael knew his choice before he ever made it.

He would follow her — wherever that led.

Even if it meant losing the stars.

Jade stood near the doorway, arms crossed, her sharp eyes softening just slightly.

"She's different now," she murmured to Arion.

He nodded once, his expression unreadable.

"She's remembering what it means to live."

"And he?" Jade asked, tilting her chin toward Kael.

Arion's gaze lingered on him. "He remembers what it means to love. And that... may be the problem."

Jade said nothing, only watched as Kael brushed his thumb over the baby's small hand — as Estelle looked up and smiled at him like the rest of the world had finally fallen away.

Arion exhaled, heavy with the weight of what was coming.

"When the time comes," he said quietly, "they'll have to choose. Between love... and destiny."

Jade's jaw tightened. "And if they choose wrong?"

Arion's answer came like a prayer to a dying star.

"Then all of us fall with them."

The Choice

They had agreed – only a few more years.

A few years before the Merge.

A few years to pretend they were human.

Jade and Arion were against their choice at first but understood. They promised to gather the remaining Starborn scattered across the globe, preparing them for the day the twin cores would rise.

And so Kael and Estelle disappeared again – a family in exile, hiding in the quiet places the world forgot.

For a while, they lived as if they were ordinary.

The cottage was small, half-swallowed by vines and wildflowers. When it rained, the leaks in the roof sang in soft rhythms that Gene learned to count along with. Estelle had filled it with small things – handwoven curtains, a chipped teapot, a blanket that smelled like rain. She liked to say the house had memory, that every wall remembered laughter.

Gene grew faster than any child should. By the time he reached his second birthday, he spoke like someone three times his age, and when he laughed, the air shimmered faintly with light. Sometimes, when he slept, stars would flicker above the house – faint, blinking dots that vanished by morning. Kael and Estelle never spoke about it. They just watched, pretending that this world – this small, fragile peace – was enough.

For Estelle, those moments felt like everything she had once been searching for – love, safety, belonging. Sometimes she'd wake in the middle of the night to the sound of Gene talking softly to the stars that weren't there, whispering as though they could still hear. She'd press her hand against his back and whisper,

"Sleep, my little star."

Kael & Estelle

Kael worked during the days, mending the valley's broken generators, teaching Gene about the earth, showing him constellations on a holographic map since the sky above had gone dark. But sometimes, when he touched the earth, he felt it tremble faintly beneath his fingers. The soil was growing colder, the river thinning.

At night, the sky bled faint streams of light like open veins not stars, but fractures. The first time it happened, Kael tried to hide it. But Estelle saw the fear behind his silence.

"Something's wrong," he admitted one night as they watched Gene sleep.

"What do you mean wrong?" she whispered.

"The planet's reacting to us. Every day we delay, it grows weaker. It wasn't meant to hold this much celestial energy. We're... contaminating it."

Estelle shook her head. "No, we're healing it. Look at him." She nodded toward Gene. "He's thriving."

Kael's eyes softened. "He's not supposed to be here either. He's half of both worlds, and both are fighting for him."

She swallowed hard. "Then let them fight. He deserves more than stars and sacrifice. He deserves the life your mother gave you."

Later that night, while Kael slept, Estelle walked through the house, letting her fingertips brush the memories etched into every corner. The first time Gene said mama – she'd been holding him by the river, the sunlight shimmering in his curls. The day he learned to write – pressing symbols into damp sand, carving his name beside hers. The night he was sick and Kael held him through until dawn, whispering the old prayers of Lyratheon to keep his fever low.

Those moments were human.

Imperfect.

Beautiful.

And for the first time in all her reincarnated lives, Estelle felt something she never had before – the desire to stay.

She whispered into the dark,

"Just a few more years."

But outside, the air shimmered – faint distortions rippling through the treetops. The stars wanted their children back.

Kael

He didn't tell her about the dream. He saw it often now – the ocean turned black, the moon splitting like glass, Lyratheon burning in the void, calling his name. In the dream, Estelle stood on the shore, holding Gene's hand. She looked at Kael with eyes that weren't human, whispering, "We waited too long."

He would wake with the taste of ash in his mouth, his skin glowing faintly like cracked marble.

That wasn't the only truth he kept from her.

Estelle moved through their days as if the world beyond their little home had ceased to exist – as if the danger hadn't followed them through every horizon. They'd changed locations six times before Gene's second birthday, each time believing it would be the last. But AstraCorps always found them.

She didn't want to see it...see that they would never stop hunting, that peace for people like them could only ever be temporary. And every night, when Kael watched her rocking Gene to sleep, a quiet dread settled in his chest. What if next time, they couldn't run fast enough? What if t his fragile world she'd built was the last one he ever got to give her?

The Fear of Losing Him

From early one, Estelle knew Gene wasn't like other children. He never got sick, never aged in slow human time. His laughter made flowers bloom early, and when he cried, thunder rolled through the hills.

At night, he'd curl against Estelle's chest, his small hand resting over her heart. He'd listen to it like a song.

"Tell me again, Mama," he'd whisper, half-asleep. "About the place we came from."

Estelle would stroke his hair, smiling softly. "Lyratheon was made of light. Every soul was a spark, every spark a story. You were one of them, long before you had a name."

"Was I bright?"

"The brightest." She said looking at her world within her reach.

He'd giggle. "Then why did we leave?"

Her throat always tightened at that part. "Because

sometimes, even stars need to learn how to dream." She'd hum to him then – the same melody that used to play in her visions, the one Kael said reminded him of their first life. Gene would fall asleep smiling, his skin faintly glowing gold against the dark. And every night, Estelle told herself: He deserves to wake up in this world, not the next.

But what she never told him was that no one really knew what would happen if they merged. Even Arion couldn't predict whether Gene, being part human and part Starborn, would survive the transition.

"He carries both worlds inside him," Arion had said before they left. "And if the merge begins... both might try to claim him."

That fear lived in Estelle's bones.

Sometimes, when Gene played outside, the air shimmered around him, starlight leaking through his skin in threads like sunlight through water.

She'd rush to him, heart pounding.

"Gene! Inside!"

He'd look up, innocent. "Did I do it again?"

She'd kneel, cupping his face. "You have to be careful, my love. Don't let anyone see."

He'd pout. "Not even the sky?"

"Especially not the sky."

After he went to bed, she'd sit by the window, watching the horizon pulse faintly where the dimensions thinned.

Each glow was a reminder: their time was running out.

And every time Kael mentioned "when we go back," she'd answer, "Not yet."

One evening, Estelle sat on the porch with Gene asleep in her lap. The air smelled of pine and damp earth. Kael was fixing the old wind turbine nearby, his silhouette framed against the last pale streaks of dusk.

She traced tiny circles on Gene's arm, watching how the faint gold beneath his skin pulsed in rhythm with her own heartbeat.

He stirred in his sleep, mumbling, "Don't go."

Her heart stuttered. "What, baby?"

"Don't go back to the sky," he whispered.

Estelle froze.

He couldn't have heard them talk about the Merge. They'd never spoken of it near him.

Her eyes welled with tears. "How do you know about that?"

Gene smiled sleepily. "The stars told me. They said if you go, I'll be alone."

Something broke quietly inside her.

She kissed his hair, whispering fiercely, "I'm not going anywhere. Not yet. Not while you still need me."

When Kael came up the steps, she looked at him through tears.

He didn't ask – he didn't have to.

The way she held Gene said everything.

That night, after putting Gene to bed, Kael found her outside, sitting in the dark.

"He dreams of the stars," he said gently. "He's part of them, Estelle. Just like us."

Her voice was soft, trembling. "I know. But he dreams of us too."

Kael sat beside her, his hand finding hers. "You're afraid if we go, he won't make it."

"I'm afraid if we go, he'll forget us," she whispered. "I want him to have birthdays. Scars on his knees. A first love. I want him to grow old."

Kael's eyes softened, full of both love and grief. "That's the one thing we can't give him."

"Then let me try," she said, voice cracking. "Let me give him a few more years to be a boy before he loses everything he holds dear..."

He looked at her for a long time – at the woman who had burned through lifetimes to find him, now clinging not to destiny, but to motherhood.

Finally, he nodded. "Two years," he said. "Then we face the stars together."

She leaned into him, whispering, "Two years. I'll make them beautiful."

But even as they loved, Earth faltered. Seasons no longer came in rhythm. The tides rose and stilled. Forests bloomed out of time, petals glowing faintly before turning to ash. Sometimes Gene would look up from his drawings, brow furrowed.

"The sky sounds sad," he'd say.

And Estelle would pretend not to hear the low hum vibrating through the air – the universe itself calling for what it had lost.

Still, she kissed him goodnight, whispered stories about fireflies and rivers and the miracle of growing up.

She knew they were lies.

But they were beautiful ones.

On the eve of their fifth year, Kael found Estelle by Gene's bedside, tracing her fingers through his hair.

"You know it's coming," he said softly.

She nodded. "I know."

He hesitated. "Will you still go, when it's time?"

Her voice broke. "If I go, will he be safe?"

Kael's eyes glowed faintly, warm and sad. "He'll be more than safe. He'll be eternal."

Estelle leaned down, pressing her lips to Gene's forehead. "Then he'll never be alone."

She turned back to Kael, her tears catching the starlight that filtered through the cracked window.

"Promise me one thing," she said. "When we go, don't look back."

Kael reached for her hand, squeezing it gently.

"Only forward. Always."

CHAPTER 16
ALWAYS

Kael stood outside one night, staring at the black sky. The air was still, heavy. Even the insects had gone silent.

Behind him, Estelle appeared in the doorway, wearing one of his shirts, her hair a dark spill around her shoulders.

"You're awake again," she said softly.

He nodded, eyes fixed upward. "It's getting louder."

"The call?"

"Yes." He exhaled slowly, the faintest trace of starlight escaping his breath. "Every night it hums through my bones. It's Lyratheon. It's waking."

She came to stand beside him, her hand finding his. "Then ignore it."

He looked at her. "You think I haven't tried?"

Her voice was small. "We have a life here, Kael. A son. A family."

He shook his head. "We have borrowed time. Every day we stay, the balance tips. You've seen it – the rivers drying, the lights dimming. Earth is dying because we haven't gone back. I know it's the farthest we've ever gotten. But we have a mission, Estelle. People counting on us to return them to their rightful home. Have you forgotten?"

She turned away, blinking back tears. "Maybe Earth's dying because of what we are. Maybe the universe doesn't need gods pretending to be people."

Kael's voice softened. "We're not gods, Estelle. We're the children of stars. And they're calling for us. Do you want Astracorps Ideaology to be true? That Starborn are detrimental to Earth? Beacause we are Estelle" he yelled.

She crossed her arms, holding herself. "And what of Gene? You'd have him grow up without us? You'd leave him to this world alone?"

"He wouldn't be alone," Kael said. "He'd be safe. He's more than us, stronger. He'll lead the others when he's ready."

She laughed bitterly. "You sound just like Arion."

He stepped closer, cupping her face gently. "And you sound like a mother who's forgotten she was born from light."

She flinched, the words cutting deeper than he meant.

"Maybe I have," she whispered. "Maybe that's what love does – it makes you want to stay."

That night, neither slept.

When dawn came, the air shimmered faintly – the first sign of the Merge approaching.

Kael stood at the window, the horizon glowing faintly gold.

Estelle sat on the edge of the bed, Gene asleep between them, his tiny chest pulsing with faint light.

Kael said quietly, "It's time."

Estelle's voice cracked. "If we go, we'll lose him."

"If we stay, we'll lose everything else."

She stood, shaking her head. "No. You don't get to choose for me, Kael. For once in all our lives – let me choose."

He looked at her, eyes filled with both love and sorrow.

"Then choose."

Her hands trembled. "I choose us. Here. Like this."

He stepped forward, his forehead touching hers. "There is no 'us' without the stars. We are fragments trying to live as whole."

Tears spilled down her cheeks. "And what if I like being broken?"

He smiled sadly. "Then let me be the piece that makes you whole again."

Outside, thunder rolled though the sky was clear – the universe itself trembling as the Merge neared.

Gene stirred, his light intensifying. "Mama...?" he murmured.

Estelle bent down, brushing his hair. "Go back to sleep, little star."

But the boy opened his eyes fully – and for the first time, they weren't human. They were galaxies.

"It's okay," he whispered, voice echoing with something ancient. "I'll see you there."

Estelle froze. "What–"

Kael's voice came low. "He knows."

The hum grew louder, shaking the house. Objects lifted off the shelves, dust spiraling in weightless rings.

Estelle's tears glittered like liquid light. "Kael, I'm scared."

He wrapped his arms around her. "So am I. But we were never meant to end here."

"Will it hurt?"

He kissed her temple. "Only for a moment."

It began quietly – not with fire, but with light.

The air shimmered as though reality itself was exhaling.

The rivers turned to mirrors. Mountains hummed. The sky bled a faint gold that deepened by the hour.

Estelle felt it first – the tremor beneath her feet, the familiar pulse thrumming in her veins. She met Kael's eyes across the room, and in that look, they both knew.

The Merge had begun.

They stepped outside. The valley that had cradled them for years was dissolving at the edges, particles lifting like dust toward the heavens.

Gene stood in the doorway, watching the world he'd known flicker and fracture. "Mama," he whispered, "the sky's waking up."

Estelle crouched beside him, cupping his face. "It's time, my little star."

He blinked back tears.

She smiled softly. "We will always love you."

Kael knelt too, his hand resting on Gene's shoulder. "You'll be safe. The others are waiting for you."

"But I want to go with you," Gene said, his voice trembling.

Kael's voice cracked. "You will. Just... in a different way."

Kael had been preparing for this moment. He found a family nearby who son adored Gene and Gene adored him. He knew in his heart he was making the right choice.

The world split open, light pouring from the cracks like molten gold. Every living thing began to shimmer – the trees, the rivers, even the wind.

Lyratheon was calling its children home.

Kael could feel its gravity pulling at him – not physical, but spiritual, like the longing of every lifetime condensed into a single moment.

He turned to Estelle. Her hair rippled in waves of light, her eyes reflecting galaxies. For the first time since she'd been reborn, she looked like who she truly was.

He reached for her hand. "It's happening."

She nodded, tears streaking her cheeks. "I know."

They stood there — light bleeding from their bodies, the air singing with energy. Gene's glow pulsed brighter, drawn toward theirs.

Arion's voice echoed faintly through the comm-link, static-filled and reverent.

"The merge has started. The twin cores are awakening."

Kael didn't answer. He was looking only at her.

The ground trembled harder. The sky cracked open, revealing the cosmos beneath — not darkness, but a river of stars, flowing like memory.

Estelle turned to Kael, her voice breaking. "Do you ever wonder what it will feel like? Being back in the stars?"

Kael smiled faintly, brushing a tear from her face. "I don't know. But I know I'll feel you there."

She laughed softly through tears. "Still such a poet, even at the end."

He shook his head. "No. Just a man who's sure of one thing."

Her breath caught. "What's that?"

He leaned closer, pressing his forehead to hers. "That we'll always be together. That I'll never leave you – or him – again."

She closed her eyes. "Always," she whispered.

He smiled, the word echoing back from his soul. "Forever."

Their light flared – gold meeting silver, human meeting celestial – until the valley itself became a sun.

The Merge took them gently, their forms dissolving into streams of brilliance that spiraled upward, weaving through the fractures in the sky.

The Earth convulsed once, then exhaled – not in death, but in rebirth.

And where they had stood, the first stars began to return.

The Rebirth

Years passed.

The Earth healed slowly – oceans deepened, forests regrew, and the night sky once again blazed with constellations. Gene grew into a man beneath that sky. He carried his mother's warmth, his father's stillness, and the faint shimmer of something otherworldly in his eyes. He became a guide for the Starborn who remained – not a soldier, not a savior, but a keeper of balance. A bridge between Earth and Lyratheon. And every night, when the world grew quiet, he would step outside and look up.

In the heart of the sky, just above the horizon, burned a new constellation – one no astronomer could explain.

It was called Lyratheon's Heart.

Three stars glowed brighter than the rest: a large one, steady and golden; a smaller, silver one beside it; and between them, a child's light – softer, but alive.

Gene smiled every time he saw it.

He knew.

They had made it home. And as the wind brushed his cheek, carrying a warmth that wasn't of this world, he swore he heard their voices – faint, distant, eternal.

"Always."

"Forever."

He whispered back, looking up at their light.

"Together."

The End

For every soul that ever loved beyond time.

If you felt the stars move with Starborn —
if you believed in love that transcends time,
in destiny written in light,
and in the kind of connection that never fades —
then this is only the beginning.

Discover more novels published with Pen of Heart:

- *The Five Kingdoms Saga — Epic fantasy. Forbidden love. Divine power.*
- *The Trapped Series — Gritty, raw urban romance where loyalty and love battle survival.*
- *Children's Titles from Pen of Heart Publishing — Heartfelt stories that bring families together.*

Join the Journey:

Follow Pen of Heart's evolving universe of stories — from galaxies to city streets, from heartbreak to healing — where every soul has a story and every love leaves a light.

Connect & Stay Updated:
Website: *www.pohpublishing.org*
Instagram: @PenOfHeartPublishing
Facebook: Pen of Heart Publishing
If Starborn touched your heart,
please consider leaving a review and sharing it with a friend who believes that love — like the stars — never truly dies.

What's Next for this author:

The Blood Moon Collection – Gothic tales of curses, immortality, and love that defies death.

The second book to Five Kingdoms: The Shattered Sea

About the Author: Breannia Venable

Breannia Venable is a passionate storyteller and avid lover of mythology. Her journey into the world of fantasy began in childhood, where she spent countless hours immersed in tales of magic, heroism, and far-off lands. This early fascination only deepened as she grew older, finding new realmsto explore through the captivating myths and legends of ancient civilizations.

Breannia earned her undergraduate degree in Communication, honing her skills in storytelling and the art of conveying powerful narratives. Her academic journey continued at Pepperdine University, where she received her Master's degree.

A pivotal moment in Breannia's life was her introduction to Greek mythology. The rich tapestry of gods, heroes, and epic quests resonated deeply with her, inspiring her to blend these timeless elements with her own creative visions. This passion for mythology and epic tales is vividly woven into her writing, where she brings to life complex characters and intricate worlds that captivate her readers.

In her debut novel, "Five Kingdoms: Rise of the King" Breannia combines her love for fantasy and mythology, creating a compelling story that explores themes of destiny, redemption, and the enduring power of unity. Through her work, she invites readers to embark on unforgettable journeys, where the boundaries of reality and myth blur, and the spirit of adventure reigns supreme.

Made in the USA
Monee, IL
04 November 2026

67381352R10148